Ether Insculped on Unbreachable Coffer and Recollation Technotronic Sanitas Capitularies

Miriam

Ether Insculped on Unbreachable Coffer and
Recollation Technotronic Sanitas Capitularies

First Edition September 2025

Written by Miriam

Contents

1 Introduction **1**

1.1 Healthcare Systems in India 1

1.2 Relevance of Information and Communication Technology(ICT) in Healthcare 3

1.3 Cloud Computing for Healthcare IT 3

1.4 Scenario of Health Information Management in Indian Hospitals 5

1.4.1 Paper based record 5

1.4.2 Electronic Medical Record 5

1.4.3 Hybrid record 6

1.5 Need for EHRs in Indian Hospitals 6

1.6 Medical terminology and coding standards for EHRs 10

1.7 Health Information Security and Privacy Regulations 11

1.8 Motivation for the study 13

1.9 Thesis Organization 13

2 Literature Review **15**

2.1 Introduction 15

2.2 Health Information System - History and Evolution 16
2.3 Health Information System - International Scenario 18
2.4 Health Information System - National Scenario 20
2.5 Cloud Computing for Healthcare 21
2.6 Security and Privacy in Managing Health Information System 22
2.7 Limitations Observed With the Existing Health Information System . . 26
2.7.1 Issues identified during field visit 27
2.8 Problem definition 28
2.9 Research objectives and scope 29

3 Development of Standardized Electronic Health Record 32

3.1 Introduction 32
3.2 Study of the Existing Healthcare Information System 33
3.3 Electronic Health Record Minimum Data Set(MDS) 37
3.4 Electronic Health Record(EHR) Standards 40
3.4.1 ICD-10 41
3.4.2 SNOMED-CT 42
3.4.3 LOINC 43
3.4.4 DICOM 44
3.4.5 NDC 45
3.5 Patient Work Flow Pattern in Proposed EHR System 46
3.6 EHR System Design 49
3.6.1 Patient demographics 51
3.6.2 Patient health history and medications 51
3.6.3 Patient clinical encounters 51
3.6.4 Legal and regulatory compliance 52
3.7 EHR Product Features 52
3.7.1 General description 52
3.7.2 Purpose 52
3.7.3 Scope 52
3.7.4 User characteristics 53
3.7.5 Specific Requirement 53
3.7.6 Security requirement 53

3.7.7 Constraints 53
3.8 Use Case Specification of EHR system 54
3.9 Overall Description of EHR System 58
3.10 Time Motion Study: Conventional and Developed EHR system 64

4 Security Framework For Cloud Based EHR System 67
4.1 Introduction 67
4.2 Proposed Security framework for EHR System 68
4.2.1 EHR system architecture 69
4.2.2 Identifying Threats 70
4.2.3 Threat Ranking 70
4.2.4 Security standards 73
4.3 Security mechanism for Secure Storage and Retrieval of EHR 74
4.3.1 Authentication Model for EHR 74
4.3.2 Authorization model for cloud based EHR 77
4.3.3 Role Based Access Control (RBAC) 80
4.3.4 Attribute Based Encryption(ABE) for secure storage of EHR system 81
4.3.5 Attribute-Based Access Control 83
4.3.6 Digital Signature 88
4.3.7 Dual Database for Securing the Patient Data 89
4.4 Secure Deployment Architecture for EHR system 92

5 Deployment of Developed EHR system in Cloud 94
5.1 Introduction 94
5.2 Deployment architecture using Public cloud infrastructure 96
5.2.1 High availability architecture for EHR model 96
5.3 Deployment architecture using Open Source cloud framework 99
5.4 Experimental Setup and Results 103
5.4.1 Launching an Instance 105
5.4.2 Accessing the Cloud 108

6 Validation of the Proposed Electronic Health Record System 110
6.1 EHR System Testing 110

6.1.1 Unit and Functional testing 111
6.1.2 System Testing 111
6.1.3 Compatibility Testing 112
6.1.4 Security Testing 112
6.2 Validation of EHR System Using Realistic Data 114
6.2.1 De-identification process in EHR system 119
6.3 Credibility factor of EHR system 120
6.4 Comparison of EHR system with other existing system 123
6.5 Case Study on EHR system 126

7 Conclusion and Future Scope 129

7.1 Conclusion 129
7.2 Scope for Future Work 131

Chapter 1

Introduction

Digitization of health records in public health facility and its instant availability in the form of electronic record any where any time health service is yet to be implemented in developing nation like India and other countries. Implementation of EHR framework for Indian health system will improve the work-flow of health services to the population. The framework for all levels of healthcare system enable efficient and continuous care to the patient. The first section of this chapter deals with the different levels of healthcare system in India. The role of Information and Communication Technology in healthcare is described in the second section followed by cloud computing for healthcare IT, Health Information Management, and different type of interoperability standards to exchange data between the various levels of healthcare systems. The chapter ends with the importance of security and privacy in healthcare and the significance of this research.

1.1 Healthcare Systems in India

Indian healthcare system is diverse, consisting of a large number of hospitals of various size and units, run by the central and state government [1]. In this system, patient care is mainly delivered through Primary/Community Healthcare Centre(PHC/CHC), Secondary level (District Hospital), and Tertiary Healthcare Facility(National level) [2] [3]. These institutions are functioning under the aegis of the Central and State Government. The private sector institutions, along with the public sector, equally participate in the provision of quality services to the needy population[4]. The different levels of healthcare system is depicted in Figure 1.1.

The Primary Healthcare Centers are the first point of contact between a healthcare institution and community. Hence, the focus is more towards maternal, and child healthcare services for the rural community, whereas secondary and tertiary care institution services are more open to the urban population.

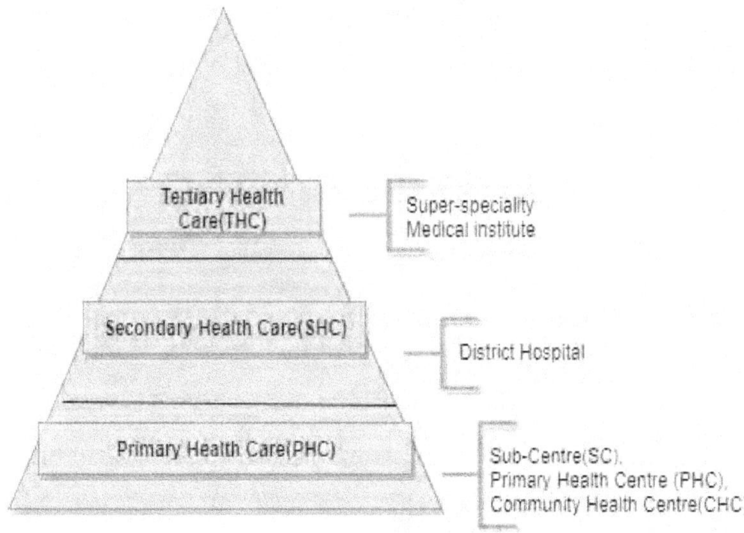

Figure 1.1: Levels of healthcare system

The secondary and tertiary care center also acts as a referral institution for primary health centers. When the patient gets referred from primary to secondary or tertiary care centers, it is always expected that the referral center should accept the case and provide quality services. But, to provide such services, the healthcare professionals expect the instant availability of the patient's complete information, as it allows them to understand the case and attain the quality outcome. But the capturing and sharing of such information becomes a challenge for the healthcare professionals and providers as most of these healthcare data gets documented on paper. Even if the health data is in electronic form, it is usually spread across multiple Hospital Information System, in different data formats and platform. The manual documentation not only consume time in capturing but also make it difficult to share patient data in an instant and real-time manner. During the last few decades, Information and Communication Technology (I&CT) has supported the healthcare professionals in managing the information related to patient care, research and education [5].

1.2 Relevance of Information and Communication Technology(ICT) in Healthcare

ICTs in healthcare delivery system have great potential to improve health of the public in both developed and developing countries by enhancing access to health information by all levels of healthcare system. Delivery of patient care is highly dependent on health data and the power of ICT mainly lies in its capacity to effectively collect, store, access, analyze and transmit patient health data electronically. ICTs have the potential to significantly contribute in preventive care, improving health service delivery, disease control, health management and research[6].

The main challenges in the healthcare delivery system is in accessing patient health information and clinical knowledge from the pool of large database. The current healthcare system is equipped with new mechanism to deal with these challenges faced by healthcare providers. The adopting of ICTs infrastructure enhances quality of healthcare delivery system [7]. Over the past few decades, the computing system has become increasingly common in healthcare setting in providing quality care and in facilitating better medical decision making. Initially, ICTs were used mainly in administrative and financial sector, the role that continue even today, but the use of the computer system in clinical decision making is growing now a days.[8]. ICTs is acknowledged as a potential solution for health care system in providing information. The new computer technology eases or enhances communication and exchange of information between all levels of healthcare system. Such technologies have the potential to save time and money, provide accurate, adequate, complete and timely health information and clinical knowledge and make it available to all levels of healthcare providers.

1.3 Cloud Computing for Healthcare IT

Healthcare ecosystem consists of doctors, physicians, nurse, pharmacist, radiologist, lab technician, and patient [9]. The data generated at the physician's office and other locations of the hospitals such as medical laboratory, medical imaging etc., during clinical encounters are typically managed, stored and maintained by the hospitals for longer period in order to provide patient care and follow-up. Due to large data, hospital sometimes finds it difficult to store and manage health data. Cloud computing has changed the way of managing all levels of healthcare data of the patient and allows healthcare professionals to access the data instantly whenever they need [10].

Cloud computing is necessary in the healthcare field for quick transfer and retrieval of the medical data during emergency in real time. Cloud computing helps in organizing the medical records at different levels of healthcare setting. Cloud can provide several benefits to all the users of the healthcare ecosystem by connecting many Health Information Management Systems together with laboratory, pharmacy, radiology system etc. Sharing of the data across different levels of the organization becomes more manageable and efficient. Cloud users of the hospital are getting benefited in terms of on-demand provisioning of hardware resources, tremendous flexibility in the usage level, significant cost reductions due to low infrastructure investments and the pay-per-use pricing model. Hospital internal IT team can decide to adopt feasible cloud strategy as public, private and community based cloud environment.

Cloud infrastructure offers the healthcare information sharing over the internet to various healthcare centers, regardless of their location and the amount of health related data[11],[12]. Most of the hospitals in India are lacking in resources to maintain and manage the patient health data. As the world moves towards digitization, one of the key challenges in developing countries like India is in making the healthcare data accessible from rural to urban in digital form. Advancement in IT in healthcare sector has made it possible to maintain and manage the patient data in digital form in all level of healthcare system. Cloud computing has emerged as a main role in providing healthcare IT solution. Therefore, healthcare organizations can largely benefit from cloud infrastructure which could be an excellent solution for the countries needs to have improved healthcare in rural areas [13].

1.4 Scenario of Health Information Management in Indian Hospitals

Healthcare levels in India follows different approaches to manage the patient health information. These are listed below.

1.4.1 Paper based record

Paper based record consists of health information about the patient which can be stored and accessed in paper format within a healthcare organization. Traditionally in India, most of the hospital manages the patient health information in paper based format, including PHC, SHC and THC level[14]. In paper based approach, the patient health information is stored in the clinical record room under lock and key. Managing paper based record requires large storage areas where the accessibility and retrieval of patient record becomes a challenge during the point of care. In the system of paper based health records, healthcare providers are unable to provide better care mainly because extremely useful patient health information is unavailable during episode of care. This is mainly due to the unstructured way of keeping the records in paper based system. The lack of healthcare IT resources are the biggest challenges for rural levels of healthcare system. In a digital world, it is very essential to take important steps in advance to ensure the availability of the right information at the right time.

1.4.2 Electronic Medical Record

EMR is an electronic version of patient health related information gathered from a single healthcare service provider. Presently most of the secondary and tertiary care hospitals have started using computer based record system for managing the health care information[15]. It allows healthcare professionals to manage medical information in a significant way compared to paper based chart. Healthcare provider can use this technology to store the data about the patient such as symptom's, treatment and prescriptions details of each visit in the hospital data repository.

1.4.3 Hybrid record

Hybrid record represents the individual patient health information in both paper based and electronic format. Today, majority of the hospital maintains and manages health records in the hybrid format[16][17]. Some of the tertiary health centers have moved from the paper-based medical records to the electronic repository, which connects different departments of tertiary health centers to a centralized data warehouse. This electronic repository allows healthcare professionals to access the patient data irrespective of departments. This system assists the healthcare professionals in receiving and sending patient data within the hospital and also controlling the flow of information at the point of care.

1.5 Need for EHRs in Indian Hospitals

With the Government of India 'Digital Health Mission', there has been a lot of advancements in the conglomerations of Healthcare and IT industry [19],[20]. This movement has led to the development of new methods and tools in maintaining the patient data in the digital form.

Each healthcare organization has a EMR system to capture patient data in an electronic format but these data are not united with each other to use it for quality decision making. Siloed vendors use different standards for clinical data exchange. With this approach healthcare user has to use various EMR applications and there is no uniform access. According to Health Information Management Systems Society's (HIMSS), [18] representation of health information system with a set of silos is as shown in Figure 1.2. In a silo system, patient records are spread across different services of the EMR application. If the healthcare professionals would like to access the complete health information about the patient, he/she needs to find the patient records form the various application by logging in multiple times to get the complete details of the patient.

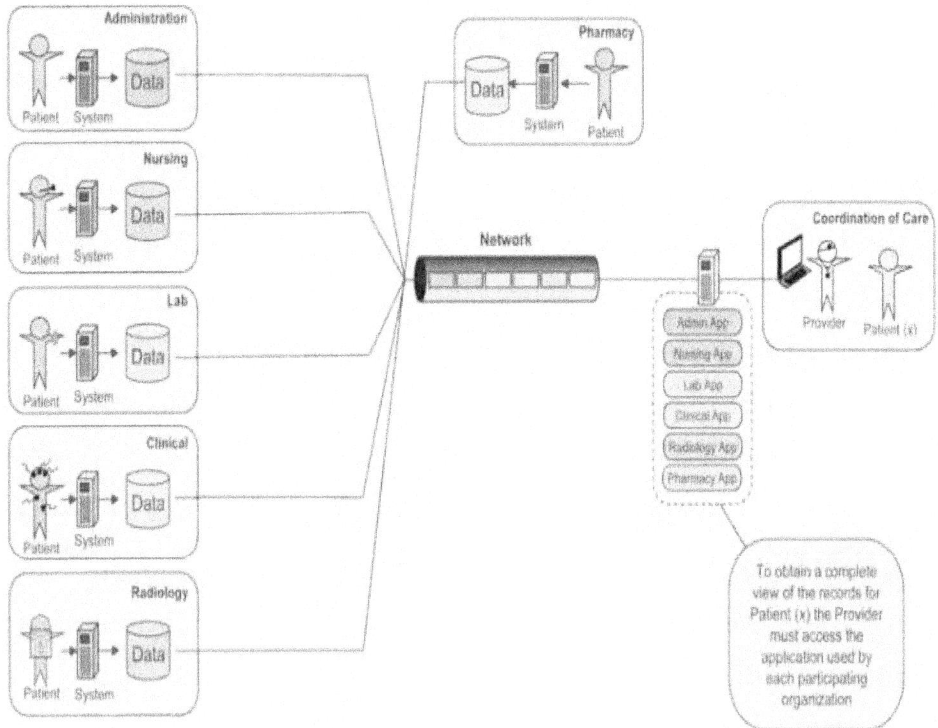

Figure 1.2: Silos Health Information System

The way of recording health information in the hospital follows the decentralized approach where all the patient data is distributed through different levels of healthcare facility. The health data of the patient is maintained and managed by the different ancillary system providers. With this approach, continuity of the care is broken and the healthcare service provider has to open and log in to the series of application to access the complete information of the patient.

There are many cloud-based Electronic Health Systems available but those are only limited to one hospital and does not have the provision to capture complete health information of patient, when patient is moving from one level of the healthcare system to other i.e., Primary to secondary to tertiary or vise-versa. The existing systems do not comply with the standards recommended by the Government of India. To boost the implementation of digital health in India, the Standards Committee of Ministry of Electronics and Information Technology (Meity) [21] has prepared a draft of National Digital Health Blueprint (NDHB)[22]. The draft was submitted to Ministry of Health and Family Welfare (MoH&FW), Government of India on 24th April 2019. Accordingly, NDHB draft was placed for the public domain for comments. The EHR is a structured digital health related data collection about individual patient.

Whenever patient visits a hospital including PHC, SHC, and THC, health related data about the patient is generated. The health information of the patient is recorded from various departments, clinical, laboratories, radiology, and pharmacy sections for administrative and management purpose. EHR system records relevant information about the patient during his/her care and it forms the legal record of care provided and it is essential for clinical communication.

With the EHR, it is possible to capture and store distinct patient health data and be made accessible at point of care including PHC, SHC and TCH. EHR contains health information about patient history, laboratory test reports, and diagnosis images stored in an digital format which is available to healthcare providers through computer network [23]. Compared to EMR, the EHR has a more comprehensive report of the patients overall health. Thus, the EHR record provides the information of various clinical encounters in chronological order to support healthcare professionals in providing quality healthcare.

The EHR is a stand-alone system, and mainly used to support patient care and clinical decision making. But to do so, EHR should be integrated to PHC, SHC and THC to make the health information available without any hindrance and barriers. To develop an integrated system, there is a need to collect the health data from the various modules including administrative, physician's entries, nursing, laboratories, radiology, pharmacy etc. The different components of the EHR system is depicted in Figure 1.3.

The first component of the EHR system is the Administrative system module, where data such as patient registration details (Patient demographic data), admission, discharge, and transfer of record from one hospital to other is maintained. The second component of the system is nursing module, where mainly apparent information of the patient is recorded such as height, weight, blood pressure, BMI values of the patient. The third component of EHR includes laboratory data such as microbiology, biochemistry etc., which plays an essential part in clinical process by furnishing relevant data to healthcare service provider. Approximately 60% to 70% of the quality decisions are decided based on the lab results.

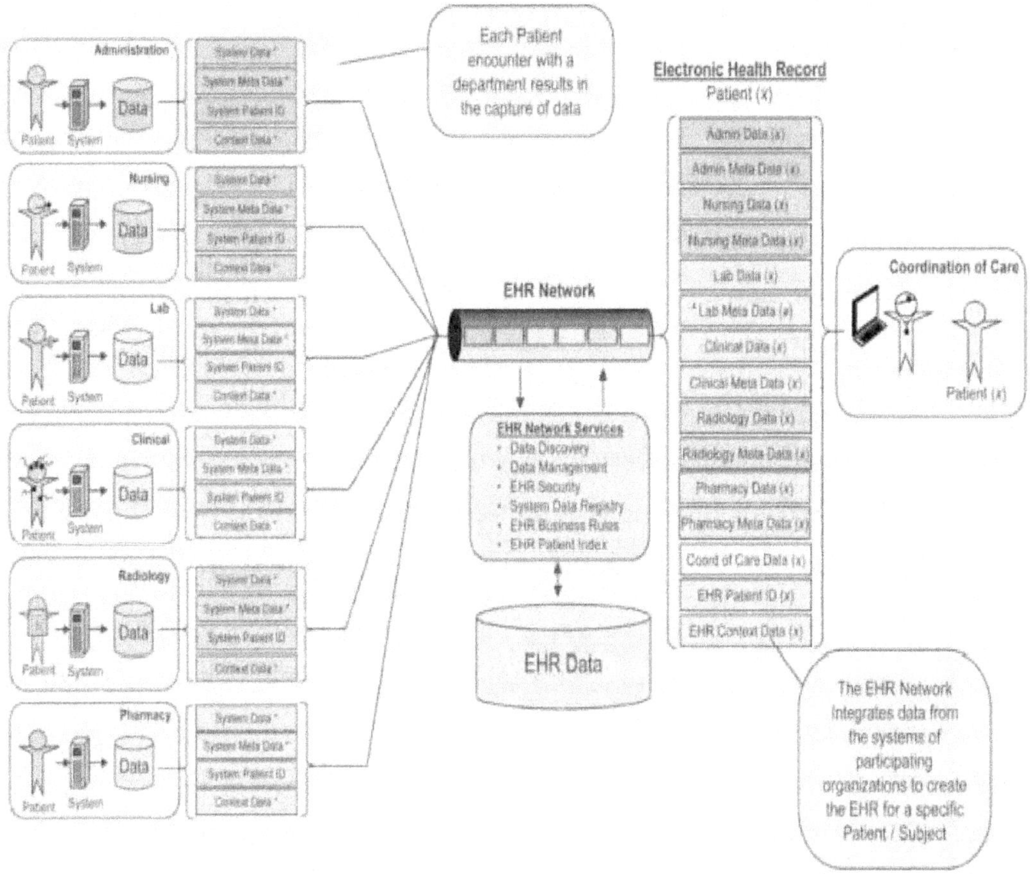

Figure 1.3: Components of EHR System

The fourth module includes radiology related data and image such as X-ray, CT Scan, and MRI. Medical imaging system called Picture Archiving Communication System (PACS) which is an imaging technology for managing digital images. The important and key component of EHR system is clinical documentation module to capture patient's clinical data such as diagnosis, procedure, complication and medication. A clinical document in the electronic form helps the healthcare provider to validate and to provide quality care to the patient. The EHR serve the purposes by providing accurate, complete, concise information of the patient, which helps next health care service provider to take quality decision. Another important module in the EHR system is pharmacy, which holds the comprehensive medication history including medication name, drug code, drug name, dosage, quantity, and allergic reaction etc., that helps in safe and effective medication to the patient.

1.6 Medical terminology and coding standards for EHRs

The primary aim of the healthcare information system is to provide lifelong clinical care to the patient at all times. To meet this requirement, syntactic and semantic interoperability of data should be maintained at all time. In India, with its out sized population spread across a large geographical landscape, to make the patient data accessible by all levels of healthcare providers is difficult as the data is not interoperable and semantic. The fundamental requirement of terminology standard is to allow effective communication, admits no doubts on healthcare users between the sender and the receiver of the healthcare information system. To represent the health concept, organized vocabularies, code sets, terminology and cataloging system is used for the communication between health information systems. The major requirements in healthcare facility is to use interoperability and standardization technique to enable easy sharing and exchange of healthcare data between the various levels. The main foundation for the interoperability is the standard terminology, which improves the effective communication between the two healthcare users [24].

Government of India has taken the initiative and formulated and published an EHR standard in September 2013 and consequently revised and published the next version on 31 December 2016. For EHR standards, Ministry of Health and Family Welfare (MoHFW) [25] suggest the healthcare facility to use the following standards namely Systematized Nomenclature of Medicine Clinical Terms (SNOMED CT) [26], International Classification of Diseases (ICD 10)[27], Logical Observation Identifiers Names and Codes(LOINC) [28], and National Drug Code(NDC) [29].

The adoption of terminologies and standard coding system in health information system ensures that data in the system is unambiguous. Content Exchange Standards(CES) are helpful to share clinical information including clinical summaries, prescriptions etc. The Digital Imaging and Communications in Medicine(DICOM) [30] and Health Level-7(HL-7)[31] are the CES used for sharing health information with other healthcare levels when patient move from one healthcare facility to other.

1.7 Health Information Security and Privacy Regulations

The aim of the hospital Health Information System(HIS) is to maintain patient data in a secure manner, considering the privacy rights of the patient and their caregivers. When the records are maintained electronically using EHR, where it has to be shared among multiple healthcare providers at a different level of the healthcare system, chances of the breach in privacy and confidentiality of patient data becomes more, which creates a threat for a healthcare institution [32]. As the medical records contain sensitive information about the patient, the breach in privacy and confidentiality lead to defamation, discrimination, and stressful conditions to the patient and caregivers.

Health details of the patient records are classified into primary and secondary record. In case of primary record, all the information are considered as confidential and hospital authority should not released it without proper consent[33]. In case of secondary record, the patient identity is hidden and patient data can be used for the research purpose like analysis, testing or prediction of the diseases. The health record system in India necessarily is safeguarded by the privacy and security standards. These standards guarantee that, health data is secure during its storage, retrieval, and transmission. For adopting security standards, the health organization must consider various aspects such as the size of the organization, complexity of the system, healthcare IT infrastructure, and application security capabilities. Security standards ensures the confidentiality, integrity, and availability of health information. The first component is the confidentiality, which identifies the rights of the patient. Every hospital is legally guaranteed to maintain the confidentiality of the patient health information. In case of any breach of confidentiality, patient can claim negligence against the physician or a hospital. Providing integrity to patient health information is essential because records are more prone to destruction or modification by an unauthorized intruder. Availability implies that the health records should be readily available to health care providers when it is required. Due to privacy concerns, the health information of the patient will be disclosed to third parties only if legislation and patient consent permits.

EHR system must also support authentication to determine that a user's requested identity is genuine. In any sector, the most commonly used authentication are passwords, digital signature, digital certificate etc. Authorization is preceded by authentication, which describes the access level of the user those who has already authenticated to the system. Access level vary from access to the entire EHR of the patient to fine-grained access of only some sections of the EHR. For instance, fine grained access definitions can be defined for users performing specific function such as order medication, admission charges, etc. Each user may have different structural and functional roles and the administration of authorization is stated as access control mechanism [34]. In cloud database server, the approved permissible activity lists are created for the each user and to access any health related data, the access control requires a validation from the created list. The Role Based Access Control (RBAC) contains set of operations with which this user can perform firm role on specific resources of the health information system.

In addition to the security requirement, EHRs are also regulated by country specific privacy laws. There are few guidelines suggested by United State of America under the Health Insurance Portability and Accountability Act (HIPAA) [35], 1996 to safeguard the patient health data that are available and maintained using Electronic Health Record. HIPAA defines privacy rules on who has authority to access to health information system and specifies the security measures including administrative, physical and technical safeguard. In India, On December 4, 2019 the Union cabinet accepted the personal data protection bill in the parliament. The purpose of this bill says that sensitive personal data can be processed only with explicit consent of the person and it prevents from the other organization to access without required consent [36]. According to Information Technology Act 2000, data which is stored, transmitted digitally can be protected by the Data Privacy Rules which mainly refers to Sensitive Personal Information (SPI) which is subset of Personal information (PI). The PI includes passwords, financial related information, psychological, physical, mental health condition, biometric, medical record and history of the patient.

1.8 Motivation for the study

The Indian healthcare system is diverse in terms of hierarchy and multitude of services provided at each level. A majority of the hospitals in India still use paper based record for health record documentation. The traditional way of health record management leads to the constant problem in providing quality of care to the patient and transfer of records continues to pose an obstacle for its success. paper based medical record lead to long queue for the registration and long wait time for treatment which can be cumbersome. Due to the unavailability of patient data on time, the healthcare professionals spend lots of time in doing the repetitive work of documenting the past medical conditions and the present history to avoid any, miss out which may affect the current wellbeing and illness of the patient. Indian healthcare system mainly in secondary and tertiary level have adopted healthcare IT systems with legacy information system in place. With this approach patient health data cannot be shared with other hospital when patient moves from one hospital to other for specialized care.

The recent trends in healthcare shows the necessity for innovative and dependable clinical information systems in a standard way for the storage and retrieval of patient health information in all levels of healthcare delivery system. To strengthen the healthcare system in India, "Digital Health Mission" has led to the development of new methods and tools in maintaining the patient data in digital form. This scenario demands for an requirement of EHR at all levels of healthcare system where data can be captured and shared on real-time at any point of care without any boundaries and obstacles. One such initiative was taken in this research to develop a Standardized Electronic Health Record for all levels of public health system. This EHR will allow the healthcare facility from Primary up to Tertiary level to capture and share patient data without any restriction and in a secure manner.

1.9 Thesis Organization

The rest of thesis is organized into different chapters as shown in the Figure 1.4. Chapter 2 discusses the literature review for the research work. The research was carried out in four phases to address the four research objectives. Chapter 3 discusses the design and development of cloud based standardized EHR and clinical and content exchange standards for public healthcare levels.

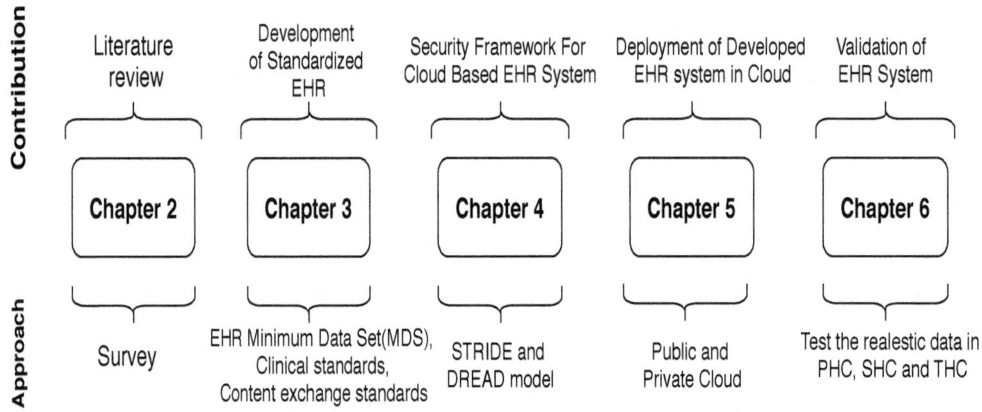

Figure 1.4: Organization of the thesis

Chapter 4 discusses the implementation of security framework for the EHR system to protect the data from unauthorized users. Chapter 5 discusses how the deployment of the developed system can be done in the cloud based environment. Chapter 6 discusses the real time data collection and validation of the EHR system. Chapter 7 contains the conclusion and future scope of the research.

Chapter 2

Literature Review

The chapter surveys the various works undertaken in Health Information System(HIS). The chapter is organized in the section wise structure as follows. Section 2.1 gives the quick introduction about the importance of Health Information System(HIS). Section 2.2 discusses about history of Health Information System(HIS) and evolution of the EHR system. Section 2.3 reviews the International scenario of the HIS. Section 2.4 reviews the national scenario of the HIS. In section 2.5 the importance of cloud computing in healthcare IT is discussed. Section 2.6 discusses the research advances in security and privacy. The research challenges and issues involved in existing HIS are listed in the Section 2.7.

2.1 Introduction

Healthcare system in India is diverse in terms of many levels(Primary, Secondary and Tertiary) and a multitude of services provided at each level. These services are provided with an aim to uplift the health status of the citizens to the satisfactory level. To provide services and reach out to the population, healthcare providers always expect that the health information of the population be available at their fingertips on a real time without any obstacles. This scenario demands for an EHR at all levels of healthcare system where patient health information can be captured and shared at any healthcare facility without any boundaries and obstacles. The implementation of the EHR should be a national goal rather than local or regional. During last few decades, I&CTs has supported the caregiver in managing the information related to patient care, research and education.

2.2 Health Information System - History and Evolution

The history of HIS[35-44] and the evolution of the EHR is shown in Figure 2.1. Medical records are an essential feature of any Hospital Management System (HMS) and it has evolved during the past several years. The latest innovations in healthcare Information Technology has changed the way of recording health information. In 1920s [37], all medical records were maintained and managed manually with the paper based documentation. Patient visit details, history, diagnosis, lab result, medication and all details were written in the paper and manually managed in the medical record room. The medical record contained key clinical information about the patient and his care. An accrediting body, Joint Commission on Accreditation of Hospitals [38] started surveying hospitals and other care facility on regular basis to check the quality of the medical care by using medical record as a tool. With these initiative greatest improvement started in the hospital for standardizing the medical record section with the defined regulations. With this goal in mind, the American College of Surgeons (ACS) [39] initiated the use of standards for the hospitals in United State and Canada, which mainly continues advances in the clinical care setting. The association of professionals exists today under the name of American Health Information Management Association (AHIMA) [40].

Medical records are documented in several ways and one such concept used for documentation of the patient health information is Problem Oriented Medical Record (POMR), which mainly focuses on specific problems that patients have. POMR [41] was introduced by Dr. Lawrence Weed in the 1950s which represents most strategic approach to document the patient care. The SOAP (Subjective, Objective, Assessment, and Plan)is a technique used to capture the patient data in POMR. SOAP improves the patient care and helps healthcare professionals to provide structured care and treatment. It clearly shows what is happening to the patient in an organized way [42].

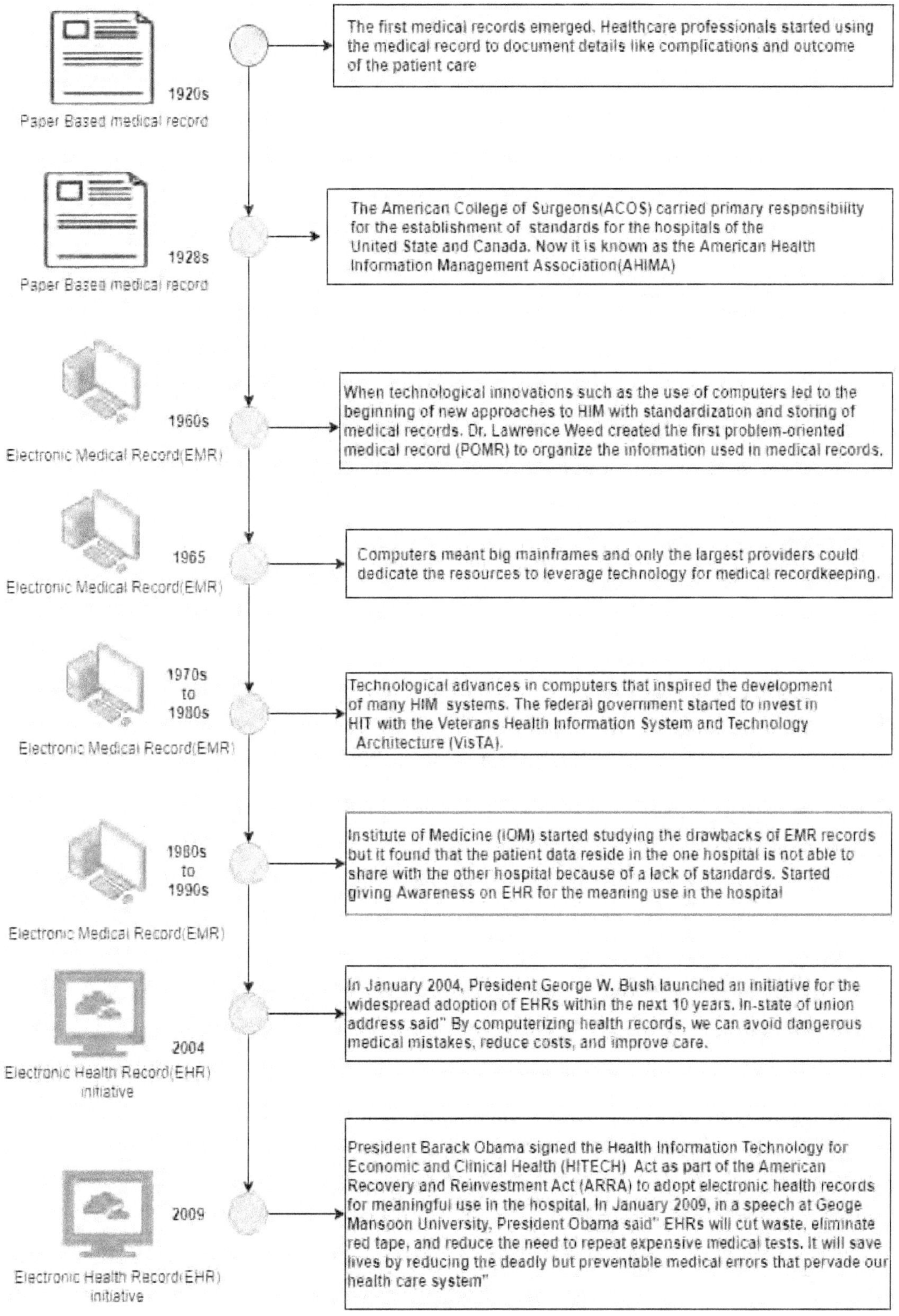

Figure 2.1: History of HIS and evolution of EHR

Advancement in the technology such as the use of computer system in healthcare sector has led to new ideal solution to the HIM for organized way of recording the medical records electronically. Electronic method of storing health information is beneficial to the organization in the technology advancement period. With increased usage of the computers in the hospital, individual department in the healthcare organization started using legacy HIS system. One such system is Electronic Medical Record System (EMR)[43] which allows transfer of health related data between the healthcare service providers of the same organization and with which healthcare professionals get quickly and easily patient data at their fingertips[44].

The major changes from EMR to EHR occur in the year 2009 during President Barack Obama term and it is named as the Health Information Technology for Economic and Clinical Health Act (HITECH Act) [45]. The act is a part of the American Recovery and Reinvestment Act of 2009 (ARRA). The prime focus of HITECH act is to motivate the implementation of the Electronic Health Records and other supporting health IT in the United Sates, so that patient can get better care [46].

2.3 Health Information System - International Scenario

The literature search evidenced the initiative taken by different countries and their implementation has become a great success in terms of improving the healthcare services by minimizing the errors in documentation and patient care. In 2001, Canada initiated to modernize its Information and Communication Technology (I&CT) infrastructure in healthcare sector by setting up objectives of standardized procurement practices, sustainable investment, and accelerate innovation adoption scale across the Canadian digital health marketplace [47] [48]. In 2015, EHRs were created for 91% of Canadians, and 91,000 clinicians were using the EHR system in their work. The EHR has benefited them in the exchange of clinical information, minimizing the errors, and improving the patient safety and overall quality of healthcare.

England initiated to modernize its healthcare system by launching the national initiative called National Plan for IT [49] [50] to provide connectivity between different levels of health records across primary, secondary, and social care by 2020. With this initiative, summary of health records were created for 54 million persons, i.e. 96 %of the England population. Using this system, healthcare professionals can monitor the summary view of the patient during episode of care.

About 90% of the private practice physicians in Germany are using EHR system [51]. With the modernization of the healthcare system, the patient data safety has been considered as the first priority by the healthcare provider during the point of care. For this specific purpose, secure health network setup is created by Federal Ministry of health Germany for transmitting sensitive medical data over the private link. In case, if the patient health information needs to be shared with other healthcare levels during that time it needs to be connected with the public network. In public network, patient can decide to hide or block any entry in their health record. But during emergency situation, the doctors can get direct access to the patient data using the secret pin provided by Federal Ministry of Health [52]. The EHR system in private sector focuses to improve the quality of services with reduced cost, and enables better data for health care management.

The survey among 501 hospitals and about 20,000 clinics of Taiwan indicated [53], [54] that the Ministry of Health and Welfare of Taiwan has initiated the National EMR Exchange Center (EEC) for sharing of record between different healthcare levels to ease the capturing and sharing of imaging reports, laboratory reports, discharge summaries, outpatient records, and outpatient medication records among health professionals and providers.

To modernize the healthcare system in Korea, nationwide [55] Health Information Exchange (HIE) platform is being built. With this initiation, most of the tertiary hospitals have started using EHRs to exchange among the healthcare facilities. The platform is connected with Application Program Interfaces (APIs) to add different services, include document registry and repository, and a master patient index.

Bahga A et al., [56] discussed about Veterans Health Information Systems and Technology Architecture (VistA) which is the traditional EHR system used in United States and covers roughly 25% of the nation's population. VistA is a collection of 168 application packages/modules. It is not a single application. They also discussed about the limitations of (VistA) with respect to the Design methodology, data interoperability, scalability and performance. Traditional health systems use the unstructured way of recording patient data, and each clinical concept is maintained as a separate table which leads to a large number of tables in the database. Also, various forms of data stored in generic structure method leads to the problem of interoperability when sharing with the other hospitals. They also discussed that the using different technical and semantic standards has resulted in conflicts in traditional EHR which ultimately lead to the problem of data integration and interoperability. Different EHR standards, different technology, and different languages also lead to the problem of exchange of data between two healthcare systems. To overcome this they have proposed two cloud based models for the United States hospital to integrate different health services together namely Open Vista and Cloud Health Information Systems Technology Architecture (CHISTAR).

In Countries like; USA, UK, Canada, Australia etc.,the EHR is developed and implemented, based on the set standards by the country to fulfill the statutory and legal requirement in delivering healthcare services to the population. For example, in US, the EHR should comply with HIPAA, ICD-9CM and Medicare and Medicaid Policy as majority of the patients are covered with health insurance. It is a mandatory legal requirement of the healthcare institution to maintain the patient health record and share in case of clinical and legal requirements.

2.4 Health Information System - National Scenario

Indian Healthcare System has also evidenced the development of ICT application but mainly limited to only few private hospitals such as Max Health[57], Apollo [58], Sankara Nethralaya [59], Fortis, etc. All India Institute of Medical Sciences (AIIMS) and the Postgraduate Institute of Medical Education and Research (PGIMER) are major public hospitals which have Electronic Medical Record (EMR) in place to share the patient data with other department of the same group of hospitals [60]. The implementation covers significant aspects from registration and billing to laboratory and clinical data. In all the hospitals, the developed EMR data are exchanged within the hospital premises and when the patient is referred to the other hospital, the patient data remains in the same hospital and patients get the print copy of their records for further consultation in hospitals, where they are referred. Also, In India most of the Health Information System does not follow the national EHR standards and always implement it based on their requirement. The need of the hour would be to implement the EHR and integrate all the levels of the public health system, where the movement of patient data becomes easier and safer during and after the point of care.

2.5 Cloud Computing for Healthcare

Cloud computing (CC) is a state-of-the-art model that provides users with real time access to a shared pool of configurable computing resources such as servers, storage, and applications [61]. Cloud computing promises to manage the traditional in-house health information technology with the help of unique features.

Sanna Sharaf et al. [62] discussed about the adoption of cloud computing in healthcare sector to fulfill the present and future demands of the healthcare services. Traditional way of recording patient health information in Hospital Information System is ineffective due to the issues including limited storage capacity, very high maintenances and operating cost. Cloud computing provides more efficient service such as computing power, software tools and applications over the internet in any place any time [63].

According to Cloud Computing Survey by G. Aceto et al. [64] the private cloud was the most popular cloud deployment scheme in the year of 2013. Although cloud computing has its advantages, it still has some issues to overcome. These issues include security of data, complete control over the cloud infrastructure, network latency issues and full access of the cloud environment. Many organizations prefer storing mission-critical data in their own infrastructure. Addition of dedicated components is possible when the cloud is owned by the organization.

Wang et al. [65] discussed about how IT can be adopted in the healthcare to automate the process flow from old technology. To fulfill the requirements of service computing and integrate the new technology, author presented a new information service model using Service Oriented Architecture (SOA). It supports reuse of healthcare resources and enables sharing of data with various healthcare facility.

Robert Birke et al. [66], discussed about how corporate data centers uses virtualization as a mainstream technology in current scenario and explained about how virtualization allows efficient and safe resource sharing in data Center. Author additionally discussed about changes in VM patterns by configuring memory and process settings of the VM.

David Freet et al. [67], proposed an effective cloud based computing service for cloud based applications. They used varieties of hypervisors such as XEN, KVM and ESX for cloud deployment model. They also analyzed the performance of hypervisor by allowing simultaneous execution of entire OS instances.

Repu Daman et al. [68], proposed an architecture for health cloud infrastructure in terms of security models. They discussed about how to protect patient data in private public cloud environment and also discussed security mechanism namely RBAC, data encryption, digital signature and time to time security audits for healthcare data.

2.6 Security and Privacy in Managing Health Information System

Research works considering various aspects related to security and privacy in cloud environment is discussed in this section.

Zhifeng Xiao et al.[69] discussed about public cloud security and privacy in terms of attribute driven classic methodology. The demand for public cloud has been increasing in the recent years. The individual and organizations tend to move towards public cloud infrastructure because of the Operating Expense(OPEX). In public cloud platform, user moves their data to cloud storage and after that user does not have control over it and only can access stored data remotely. So, information security is a foremost problem in public cloud storage. Preservation of confidentiality, integrity, and availability is the most critical element of cloud computing.

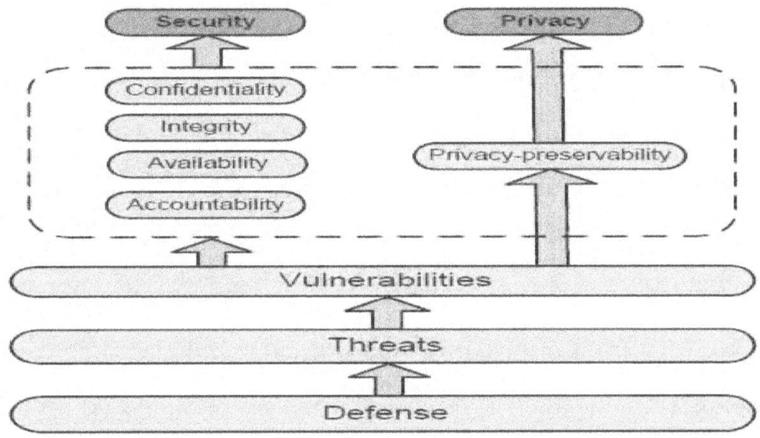

Figure 2.2: Ecosystem of cloud security and privacy

Requirements for data security given by the authors is shown in the Figure 2.2 and they stated that depending on the service model it various. Author introduced defense technique to the unsolved security issues, vulnerabilities and threats for the cloud computing. Using all these defense mechanism and solution finally cloud user can come to a conclusion that public cloud is safe to store the sensitive data on the cloud providers storage.

Clemens Scott Kruse et al. [70] have given descriptive investigation on patient privacy and security for healthcare industries, if they adopt Electronic Health Records. They have mentioned about the specific security methods and techniques to protect the sensitive data of the Health Information System(HIS). With the survey of 25 top journals on security and privacy, security techniques are categorized into administrative, physical, and technical safeguards. In addition, authors have suggested the need for advanced security technique for a EHR system to protect sensitive nature of the data.

Amer K [71] presented a conceptual framework to overcome the complex security issues in the HIS. There is no completely secure system in place to safeguard the HIS. The major goal of their framework is to use the health records in a safe and ethical way in various levels of healthcare system.

Choi Young B. et al. [72] have assessed the security challenges associated with the health information based on HIPAA and security rule. They have done a detailed study on various techniques of creating, receiving, maintaining, or transmitting health information in the modern healthcare system and suggested all healthcare organizations to comply with the new HIPAA regulations to secure the data.

Almulhem et al. [73] proposed client server model of healthcare system to analyze the attacks inside the healthcare organization. Authors discussed an attack tree (Threat modeling methodology) tool which can be used to analyze the attack which affect the healthcare system. Their system guarantee that the health record is securely protected from attacks. Authors also performed qualitative and quantitative analysis for security countermeasures which greatly reduces the attack on the system.

Fernandez Aleman et al. [74] mainly focuses on privacy and security in healthcare data when it is in transit. Healthcare data is prone to attack when sharing outside the hospital premises. Authors suggested the use of the following standards and regulations: the HIPAA regulations and European Data Protection Directive 95/46/EC for protecting the health-care data. Additionally, authors suggested the use of a symmetric key, asymmetric key, digital signature, Role-Based Access, and audit log for preventing unauthorized access of the health record system.

Ardagna, Claudio A., et al. [75] have proposed an access control solution which aims to regulate the break-the-glass exemptions that efficiently arise in all levels of healthcare systems. Break-the-glass is an approach by which a person can gain access to certain restricted data during emergencies. However, this allows malicious users to gain access to the privileged information by manipulating the security principle.

Anitha K et al. [76] have proposed a Data storage Lock algorithm to maintain confidentiality of patient data when it moves to cloud storage. To gain access to the storage resources, cryptographic key called private key (pk) standard is used. All the cryptographic keys and index tables are maintained and managed in the cloud service provider end and patient health data gets accessed only by authorized users with assigned private key.

Jieun Eom et al. [77] mainly focuses on patient controlled attribute-based encryption, in which patient has complete command over the health record. In their method, patient has complete rights to give access to any healthcare professional. The model enables the patients to control access to their health data stored in the database. When a patient moves to another hospital, sharing data becomes easy and eliminates redundant work. Additionally, the authors have also implemented services during emergency situation, where patient health data is available to the doctor with special permission.

There are few guidelines suggested under the HIPAA, 1996 to safeguard the patient health data that are available and maintained using digital record. To make the healthcare system to run smoothly, HIPAA has suggested a set of valuable security and privacy guidelines within the purview of federal law [78] [79]. The essential guidelines related to privacy, confidentiality, and security are presented in Table 2.1.

Table 2.1: Security and privacy requirement as suggested by HIPAA

Requirement	Description
Patient's interpretation	This indicates that patient is the owner of the health records and he/she has the right to know about the content of the record which is managed by the healthcare provider.
Patient's control	The control permits patients to give access to their own records with the access policy.
Confidentiality	Confidentiality indicates that healthcare data should be private and it must not have provision to give access to the unauthorized users.
Data integrity	The data integrity guarantees that shared or stored information must be a real presentation of original message and without any modification or revision.
Consent Exception	Patient's health information can be accessed only during emergency cases without his consent.
Non-repudiation	All the patient's health information is recorded and stored with supporting evidence to avoid disputes or distrust.
Auditing	The health data of the patient should be monitored all the time to ensure the data on transit and rest, both are secure and protected

There have been extensive research studies conducted in the areas of cloud data access and storage system. An integrated use of a Discretionary Access Control (DAC) framework [80] along with RBAC has been introduced to safeguard patient data to ensure confidentiality from intruders. In [81], a Trust-Aware Role Based Access Control (TA-RBAC) model has been proposed which takes into consideration the trustworthiness of the users before providing them access to certain information. The trust factor incorporated in the system allows an easier way to assign roles and deny read-write operations if the respective user fails to meet the threshold trust requirements [82]. Different methods of encryptions including Attribute Based Encryption(ABE), proxy re-encryption and lazy re-encryption are combined by Yu, Wang, Ren and Lu in order to attain user access policy, confidentiality and secret key accountability of the users [83]. In [84] V Koufi et al. implemented ubiquitous access control methodology to access emergency medical services during emergency situation. In addition to access control systems, efforts have also been made to ensure that records have been stored after encryption and that data is transferred over a secure connection. Zhifeng Xiao et al. [85] described about administrative and technical safeguard. Using administrative safeguards protects patient data through malicious emails by preventing access to unauthorized users. The technical safeguard ensures the protection of health information from the unauthorized access using proper access controls mechanism.

2.7 Limitations Observed With the Existing Health Information System

From the literature survey certain limitations in the existing health record systems are identified. These limitations are addressed during the development of the proposed EHR system.

- Availability of health information in all levels: Patient health information plays a vital role in decision making but the existing system fails to provide any evidence of capturing the complete demographic information and clinical details of the outpatient during the continuity of the care.
- Uniform database design: Database architecture of current health information system is designed to support single clinical application and does not allow any modification to meet the requirement of the uniform accessibility of data across different levels of healthcare system. This kind of design makes poor data collection, results in data exchange difficult between various levels of healthcare facility. The content of the HIS system must be well organized and system should allow instant patient data access during point of care. The interface of health information system should be uniform throughout the various levels(Primary, Secondary and Tertiary) of healthcare system.
- Lack of standards: Representation of data type in all the health-care levels should be unique including clinical and non-clinical information. Existing health information system is being developed by different vendors by using different programming languages and database. Hence achieving interoperability is a challenging task in the healthcare level. Now a days various representation have

been used, but to harmonize this into a consistent representation healthcare standards is necessary.

- Security and Privacy: Health Information System contains sensitive information about the patient like demographic, specific diagnosis, allergies, and procedures. The massive importance of this sensitive information makes HIM an attractive target for the attacker and hackers. Due to this reason, HIS poses new challenges and threat to the patient privacy and security. However, this risk can be mitigated by developing suitable model with standard security and privacy regulations when storing patient data in semi trusted servers, which ensure patient-centric privacy control over their own HIS.

2.7.1 Issues identified during field visit

Various healthcare centers have been identified and visited to understand the patient flow in the centers and to understand the process of recording the patient's information. The main purpose of the visit is to understand the significance of EHR system at all levels of hospitals. Following are the government primary healthcare centers visited during December 2016 to February 2017.

- Sub center: Moodabettu, Udupi
- Primary Health Center: Guddeangadi-Padukone
- Community Health Center: Byndoor

To understand the process of health information system among the above mentioned healthcare facility, a validated survey tool has been used to collect the data. The healthcare staff, present during the time of visit have been interviewed and asked on how the existing health information system is helping them in management of patient data and sharing of data to the referral facility. During visit, the working of existing health information system is also observed and the major findings are noted down.

The major issues identified during the visit are mentioned hereunder that have been addressed while developing the proposed Electronic Health Record.

1. Infrastructure to support EHR.

2. Security concern of patient's critical information.

3. Concerns about storage and accessibility of information.

4. Integration of disease related awareness in local language for EHR.

However the main focus of research is to design and develop a prototype for Secure EHR system. The system will address the first, second and third issues. The fourth issue is not in the scope of the research.

2.8 Problem definition

Current healthcare system highly depends on the instant access to the relevant patient health information during the point of care in achieving quality healthcare. Non availability or delay in receiving these necessary information leads to medical error, harm, and suffering to patient. Most of the rural and urban hospitals in India are managing the patient health information in different record format including paper-based, Electronic Medical Record and Hybrid records. When patients are transferred between different levels of care facility that leads to the incomplete information to the health care professionals. Sometimes patient needs to undergo same test again because doctors at the referral level cannot access the previous lab test which is done at other care facility. During the process of treatment the healthcare professionals always looks for new way of recording the patient health information where standard way of recording improves the accessing speed of the record.

The observation of existing health information system in the selected healthcare facility and the feedback received from the healthcare professionals during the visits also evidenced that the healthcare professionals spend more time in searching for adequate and complete health information of the patient and less time in providing patient care and carrying out other healthcare activity. The current challenge is to produce a standardized EHR system. Thus, patient record must be made available anywhere within the hospital/clinic, at home, outreach areas or any part of the globe.

When the records are maintained electronically using EHR, where it has to be shared among multiple healthcare providers at a different level of the healthcare system, chances of the breach in privacy and confidentiality of patient data becomes more, which creates a threat for a healthcare institution. As the medical records contain sensitive information about the patient, the breach in privacy and confidentiality lead to defamation, discrimination, and stressful conditions to the patient and caregivers.

The issues related to the current practice provided the scope of research where a single standardized cloud based EHR web application can be prescribed to the end user (Healthcare Stakeholders) addressing all the issues with the current practice and completely assist the primary, secondary and Tertiary care public healthcare professionals with complete patient health information for quality decision making during the episode of the care. The ultimate goal of this research is to completely support healthcare professionals of the different healthcare levels with required health information of the patient at all the time when it necessary.

2.9 Research objectives and scope

The EHR is a digital version of patient health information. EHR collects health related data about individual patient in a systematic way. Currently the health records have been represented in different formats which includes unstructured, narrative text or as structured database within proprietary Hospital Information System. The current healthcare environment consists of a set of loosely connected and organizationally independent units. Health information details of the individual patient are dispersed across many levels of the health care system in which standardization of health record and consistent data availability are the ultimate and important requirements to benefit the functioning of health care system.

The research objectives has been formulated based on the literature survey, field survey and from the problem statement to successfully accomplish the aim of designing and developing a standardized EHR for public healthcare sectors in India. **The objectives included are:**

1. **Objective 1 : Development of Standardized Electronic Health Record System**

   Rationale:

   - To learn and understand the features and limitations with the existing system and address the issues during the design and implementation of the web based EHR system.

- To develop standardized EHR system by understanding the contents of the medical documentation as per the regulation set by the Govt of India. EHR is designed and developed as per the Minimum Data Set(MDS) and by using health-care standard codes such as SNOMED CT, ICD- 10, LOINC, DICOM and National Drug Code(NDC).
- To improve the accessibility and availability in all three levels of patient health information.

2. **Objective 2: Develop a suitable approach for secured storage of Electronic Health Records**

   Rationale: To guarantee that the health information of the patient record stored in the cloud database is secure. For this security and privacy standards are to be adopted. This ensures atomicity, consistency, and durability of the record. These standards guarantee that, health data is secure during its storage, retrieval, and transmission. For adopting security and privacy standards, a security framework for EHR system is proposed. The framework includes physical, administrative, and technical safeguards which ensures the confidentiality, integrity, and availability of EHR system. Threat modeling tools are used to identify and mitigate the potential security issues.

3. **Objective 3: Develop a suitable access control mechanism for retrieval of Electronic Health Records**

   Rationale: To implement access control mechanism in all levels of healthcare system, which allows authorized healthcare professionals with valid authentication to access patient data without any difficulty. When the patient gets referred from primary to secondary or tertiary level for specialized treatment or consultation, the healthcare professionals at the referral level will have access to patient's health record.

4. **Objective 4: Deployment of the developed system in Cloud Environment**

   Rationale: To deploy cloud based EHR web application in a public and private cloud setup to test the availability of the patient health information in various levels of healthcare system. The cloud based EHR application allows the healthcare facility from primary up to tertiary level to capture and share patient data without restriction and in a secure manner.

Summary

Literature related to the history and evolution of Health Information Systems(HIM) is discussed. The global and national scenario of the existing EHR systems of various countries and the privacy guidelines specified for EHRs are explained. The literature review helped to define the scope of the research work. The subsequent chapter discusses the detailed methodology adopted for developing standardized cloud based application.

Chapter 3

Development of Standardized Electronic Health Record

EHR is a digital version of patient health records. It can present clinical information and complete patient health data to the healthcare professionals during the point of care. Standardization in EHR improves the easy sharing of health information between the various levels of healthcare system. The availability of the patient data in a timely fashion can contribute the improvement of patient information and performance of the Health Information System. This chapter deals with a detailed methodology followed for the development of standardized EHR in all levels of healthcare system. Figure 3.1 represents the different sections of the chapter to develop a standardized EHR in Indian healthcare scenario.

3.1 Introduction

The Government of India 'Digital Health Mission' [86] has led to the development of latest methods and tools in maintaining the patient data in the digital form. These initiatives by MoHFW has guided the healthcare institution to develop EHR to streamline the process of managing health information and support the healthcare system in improving population health. The work proposed is to develop a standardized EHR. The EHR is one such solution to support the healthcare facility, irrespective of levels and sizes to improve patient care by eliminating the paper based medical records. In order to reach the research goal, the public healthcare facilities of

Udupi district were visited to understand the patient flow and to find what type of activities are initiated to adopt digital health records in all three levels (PHC,SHC,THC) of healthcare system. This EHR will allow the healthcare facility from Primary up to Tertiary level to capture and share patient data without any restrictions and in a secure manner.

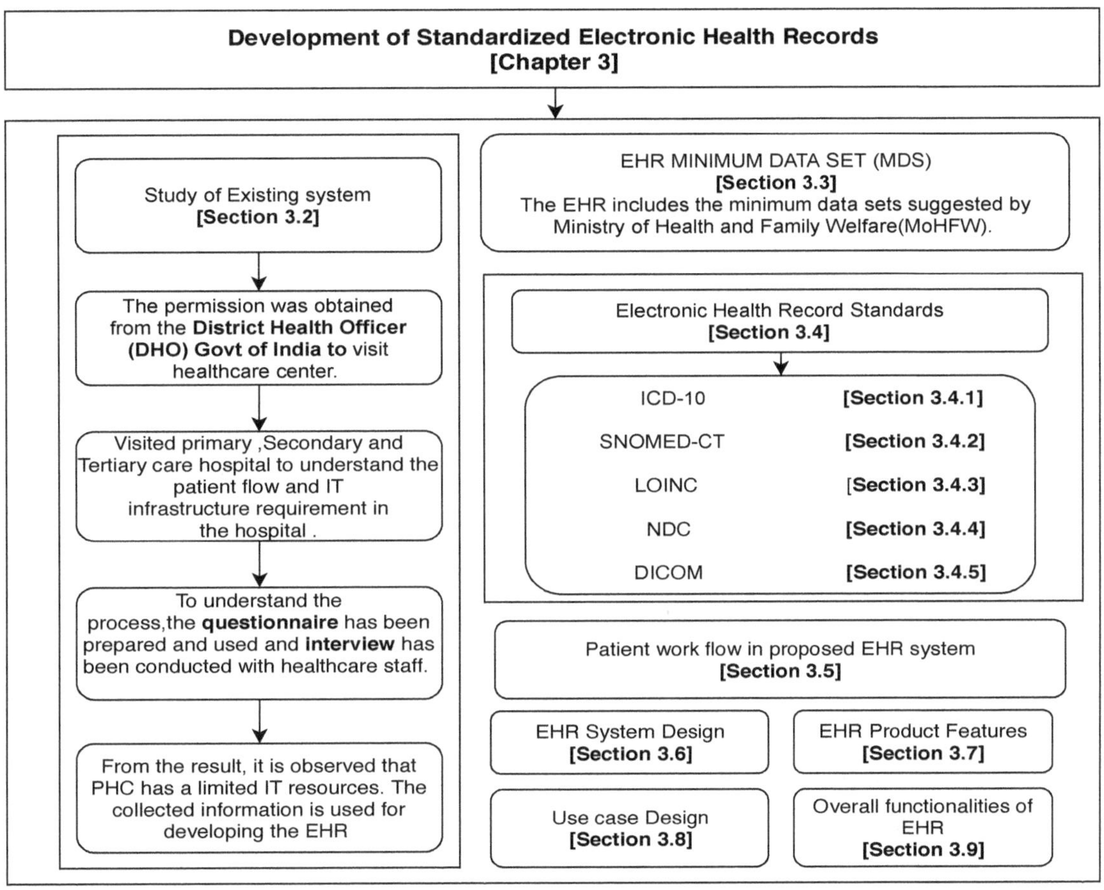

Figure 3.1: Chapter organization: EHR system development

3.2 Study of the Existing Healthcare Information System

To design and develop a new IT application requires assessment of the utilization pattern of the existing system among the end users. The data on the assessment helps in understanding the end users satisfaction level with the existing system and assist the planner in planning of better system. The public healthcare facility has been identified and the questionnaires was given to the healthcare professionals to understand the significance of digital based system at all levels of healthcare facility. The permission was taken from the District Health Officer (DHO) to visit the various healthcare centers. Following are the government primary healthcare centers visited during December 2016 to February 2017.

- Sub Center: Moodbettu, Udupi
- Primary Health Center: Guddeangadi-Padukone
- Community Health Center: Byndoor

To understand the process, the questionnaire has been prepared and used and interviews have been conducted with healthcare staff. Firstly, the respondents were briefed with the research and its objectives and the reason of the study. Based on the questionnaire the issues have been recorded and considered into the research to build the EHR system.

Table 3.1: Distribution and use of information storage methods in healthcare system

Patient Health Information Management	**Paper Based Record**	**Hybrid Record (Paper+Digital)**	**Fully Digital Record**
Demographic Details	✓	✗	✗
Physician and nurse notes	✓	✗	✗
Patient history	✓	✗	✗
Clinical exam observation	✓	✗	✗
Diagnosis details	✓	✗	✗
Treatment procedure	✓	✗	✗
Lab test details	✓	✓	✗
Medication details	✓	✗	✗
Medical Images	✓	✓	✗

The Table 3.1 presents the result of the survey conducted among the public health care facility to know about the present Health Information System. The result of the study indicated that Primary levels healthcare such as Sub Center(SC),Primary Healthcare Center(PHC) and Community Health Centers(CHC) are managing patient health information in the paper based record. For managing laboratory result and medical imaging they are using hybrid record which is a mixture of both paper based and digital.

During public health level visit it is observed that when patient visits the hospital, a token number is issued. The patient must wait till their turn comes and later healthcare professionals records vital sign and diagnosis details of the patient in a case sheet book. During the next visit to the hospital the patients are required to bring the case sheet book for updating the health information. The physician updates the case sheet during every visit with the following information including symptoms finding details, investigation result, medications details etc. In PHCs patient's case sheet was the only the record to give the details about the health information. The case sheet is the paper based physical copy and it was vulnerable to damage and over time reduce its quality. This is the only one way of portability solution if the patient moves to the different types of healthcare levels.

Medical documentation at the PHC level is mainly a paper-based and they maintain registration book, examination book, and the treatment book. The five year old active patient records are maintained in the front desk of the medical record department(MRD) for faster access. All other records are maintained in the repository which are still available for future reference. Healthcare facility staff needs to spend enormous amount of time and effort during the point of care and it becomes a cumbersome process. The solution could be to maintain an automated electronic system, preferably EHR, to capture patient and community health data to automate and streamline healthcare service provider work flow. The EHRs can facilitate clinical alerts to guide the primary care physicians about the diagnosis and treatment of the patients in real-time.

During the next visit to PHCs, the IT-infrastructure facility used for managing and maintaining the patient information was assessed and is depicted in Table 3.2 and Figure 3.2. It shows the present infrastructure facility in PHC levels.

Table 3.2: Infrastructure Facility in PHC levels

Levels and Facilty	**Total computer facility**	**Router facility (Broadband)**	**Printer**	**Scanner**
Sub Centre(SC)	0	0	0	0
Primary Health Centre(PHC)	1	1	1	1
Community Health Centre (CHC)	2	1	2	2

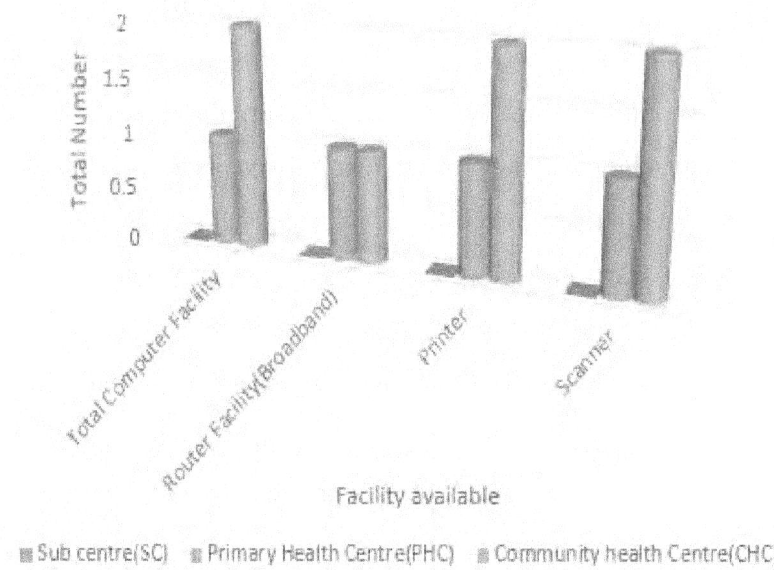

Figure 3.2: Infrastructure facility at PHC

To implement IT application infrastructure in healthcare system especially in rural places, the main obstacle is requirement of the high speed Internet connectivity. To address this national level problem, the government of India has proposed a new project BharatNet for providing high-speed broadband optical connectivity to all rural areas. With this communication infrastructure all rural places in India will get 1 Gbps (Gigabyte per second) bandwidth capacity at rural level setting. Additionally,the government requested Telecommunications(Telecom) company to list all unconnected villages in India and have planned to connect with Telecom services by 2020.

To understand the way of managing the health record at Secondary and Tertiary level, District hospital, Udupi and Kasturba Hospital, Manipal were visited. During the visit, it is observed that recording patient health information is done with the hybrid record structure which is combination of both paper and electronic record. Most of the medical records are paper based and some patient data are maintained on the computer. These include registration information, administrative, and billing data. Many of the document included in a paper based record consists of demographic data, vital signs, patient health histories, diagnosis details, medications, allergies, lab test result and medical imaging etc. As technology advanced, few healthcare facility in India have started implementing Electronic Medical Record(EMR). Often these records are not integrated with other healthcare facility and they are recorded and stored in the HIS system which is developed using proprietary software. When the patient is referred to the other hospital, the patient data remains in the same hospital.

During assessment of the existing system in practice, it was observed clearly that the system partially assists them in documenting the patient information, patient care, and reporting but does not support sharing of health information between the different levels of healthcare facility. Most of the respondents felt that the development of the EHR application reduces the hospital cost in the long run. For healthcare staff, it reduces their workload in terms of documentation and reporting. It also increases the productivity of the various levels of healthcare facility. The respondent expected that the proposed system should support them in clinical decision making through the instant access of the clinical standards, practice guidelines and patient information. They also commented that the healthcare application should be user friendly. It should have the features for the easy documentation, report generation and also assist in improving the clinical outcome of the patient. Considering the rural and urban healthcare system, the way of managing the patient health information in mind, development of standardized EHR has been proposed with the objectives to reduce healthcare professionals time in information management and retrieval, and enhance their productivity and performance by assisting them with rich source of data in standard coding system. The need of the hour would be to implement the EHR and integrate various levels of public health facility, where the movement of patient data becomes easier and safer during and after the point of care.

3.3 Electronic Health Record Minimum Data Set(MDS)

To design and develop EHR, MoHFW, Government of India has proposed EHR Minimum Data Set (MDS). The MDS ensures that health information of the patient is precise, unambiguous and acceptable to all health stakeholders. Using this digital form of data, patient record can be accessed at any place, any time and at any hospital. The captured data from the EHR can be used for diagnosis and presenting the health status of population.

Table 3.3: Dataset for Patient Registration

Data Item	Data Type	Data Length
UHID	Numeric	12
Alternate UHID	Any	Any
Patient name	Alphanumeric	Sufficiently Large
Date of Birth	Date	Fixed
Age	Numeric	Fixed
Gender	Alphanumeric	1
Occupation	Alphanumeric	Sufficiently Large
Address Type	Alphanumeric	9
Address Line 1	Alphanumeric	Sufficiently Large
City	Alphanumeric	Sufficiently Large
State	Alphanumeric	Sufficiently Large
Pin-code	Alphanumeric	Sufficiently Large
Patient Phone Number	Alphanumeric	16
Patient Email Id	Alphanumeric	255
Medical Emergency(ME) contact person Name	Alphanumeric	Sufficiently Large
ME Contact Person Relationship	Alphanumeric	9
ME Contact Person Phone Number	Alphanumeric	16

Table 3.3 shows the set of data items and its corresponding data types with respect to the patient registration process. This dataset is used for the development of EHR for the registration process.

Table 3.4 shows the set of data items and its corresponding data types used for recording health details about the patient. The nurse enters all these details into the EHR system. The developed system has strictly followed these data sets in the respective database design.

Table 3.4: Dataset for recording clinical details of the patient

Data Item	Data Type	Data Length
Clinical Examine Height(cms)	Floating	3 digits,2 decimals
Clincal Examine Weight(kg)	Floating	3 digits, 2 decimals
Clincal Examine Temperature(.C)	Floating	2digits,2 decimal
Clinical examine vitals systolic BP	Numeric	3
Clinical examine vitals systolic BP	Numeric	3
Clinical examine pulse rate	Numeric	3

Table 3.5 shows the data items and its corresponding data types used for the patient treatment process. The doctor fills the details of the patient as per the defined MDS data items to the EHR system.It also stores the patient history details. The details are filled by the doctor at the time of diagnosis.

Table 3.5: Dataset for Patient Diagnosis and treatment process

Data Item	Data Type	Data Length
Reason for visit	Alphanumeric	4096
Encounter Number	Numeric	4
Patient past history	Alphanumeric	4096
Patient family history	Alphanumeric	4096
Patient personal history	Alphanumeric	4096
Allergy status	Alphanumeric	4096
Blood group	Alphanumeric	3
Clinical observation	Alphanumeric	4096
Investigation Result	Alphanumeric	4096
Diagnosis name with ICD	Alphanumeric	Sufficiently Large
Diagnosis Code with ICD	Alphanumeric	Sufficiently Large
Diagnosis(Details)	Alphanumeric	4096
Treatment,Plan, and Procedure	Alphanumeric	4096

Table 3.6 shows the storage format of patient medication details. The details are filled in the User Interface(UI) by the doctor as per the guidelines mentioned in the MDS medication format.

Table 3.6: Dataset for medication(Drug) details

Data Item	Data Type	Data Length
Drug Name	Alphanumeric	Sufficiently Large
Drug Code using NDC	Alphanumeric	Sufficiently Large
Strength of the Drug	Alphanumeric	Sufficiently Large
Dose of the Drug	Alphanumeric	Sufficiently Large
Route of the Drug	Alphanumeric	Sufficiently Large

3.4 Electronic Health Record(EHR) Standards

Electronic way of recording patient health information has grown progressively in recent years. The patient data present in digital format is stored in the individual machine and individual care delivery setting. These distinct island of information could hinder critical sharing of data between the various levels of healthcare facility. Standards play an major role in improving the interoperability of HIS. Thus standard coding provides easy sharing of health information, reduces uncertainty, improves work flow and accurate analysis of healthcare data.

The primary aim of the healthcare information system is to deliver lifelong clinical care at all times. To meet this requirement, syntactic and semantic interoperability of data should be maintained at all times. In a country like India, with its large population spread across a large geographical landscape, to make the patient data accessible by all levels of healthcare providers is difficult as the data is not interoperable and semantic. To meet consistent representation of the interoperability requirements, standardizing vocabularies, or mapping between different vocabularies are necessary. Including the standard terminologies and codes in computer processable presentation into a HIS, support automation and ensure that data is in an analyzable state at all times. The different standards incorporated in the developed system are depicted in the Figure 3.3.

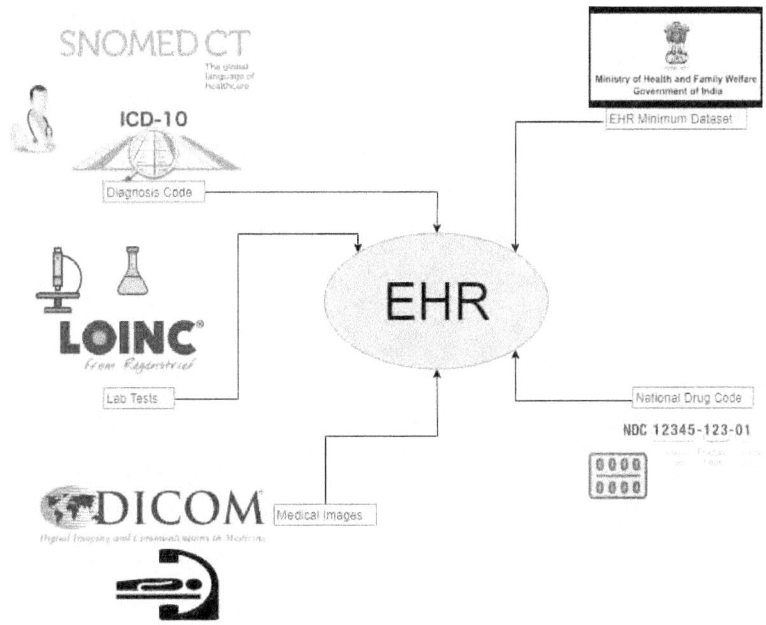

Figure 3.3: Electronic Health Record Standards

3.4.1 ICD-10

ICD-10 is the 10th version of medical classification list by the World Health Organization(WHO) [87]. The ICD stands for International Classification of Diseases(ICD) is a coding standard for disease, health management, diagnosis and clinical purpose. ICD is available in the different flavors such as ICD 9, ICD-10, ICD-10-CM(Clinical Modification) and the WHO is currently working on the 11th revision of ICD called ICD-11. ICD 9 code set contains 3-5 characters in length. First three character is used for categories of codes, further two digits gives the details about diagnosis. ICD 9 is the older version of the clinical standards and there is no provision to add new categories of diagnosis procedure codes. The existing codes in ICD 9 are not descriptive enough to give details to the healthcare providers.

The ICD 10 is a widely recognized international system for recording diagnoses. By using ICD 10 codes, the patient data will be consistent when describing diseases, morbidity, and mortality. ICD-10 code set is having three, four, five, six or seven characters. First three character is used for categories of codes. Further expansion gives the details of diagnosis. The usage of ICD 10 codes for the patient A is described with an example use-case. Patient A is diagnosed for fracture in his/her wrist and doctor has initially captured data using an older version of ICD codes in which, it does not differentiate between right and left wrist and gives fewer details for the doctors. A newer coding system ICD 10 is flexible enough to incorporate emerging diagnosis data. Also, ICD 10 differentiates between right and left wrist of the patient as clearly shown in Figure 3.4.

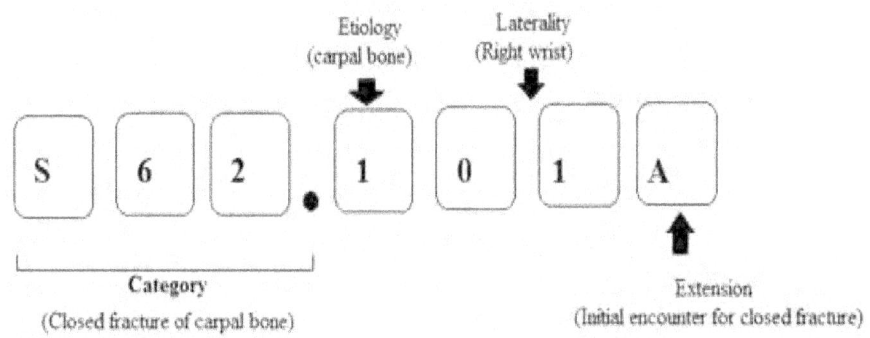

Figure 3.4: ICD 10 Increased specificity example

3.4.2 SNOMED-CT

SNOMED-Clinical Terms is a computer processable, comprehensive, multilingual clinical terminology. The main purpose of the SNOMED-Clinical Term is to standardize the way of recording of the patient data in healthcare terminology. It has no limits on the codes that can be used compared to ICD-10. It is having the comprehensive clinical scope and covers the procedure, diseases, and clinical data. Using this coding standard, patient health status such as problem list, allergies and diagnosis, and rules are stored in a consistent way across the site of care. By standardizing, the terminology allows clinically relevant meaning based retrieval and allows more accurate reporting of data. SNOMED CT is a semantically rich, clinically validated, controlled vocabulary that simplifies evolutionary growth in expressivity to meet evolving requirements [88].

The SNOMED CT can easily be integrated with an EHR system for better user interface and is simple to capture the clinical ideas and share the information across healthcare facilities. SNOMED CT has been divided into 19 hierarchy. Every code concept is linked to related concepts (attribute relationships). An example for viral pneumonia using SNOMED-CT is shown in 3.5.

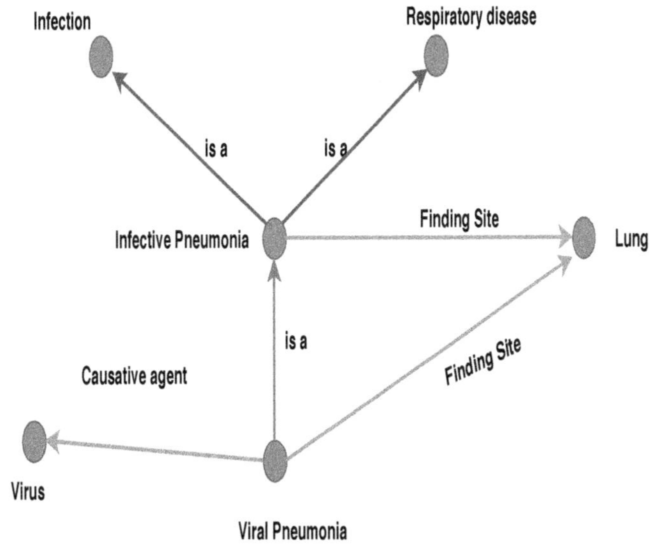

Figure 3.5: An example for viral pneumonia using SNOMED-CT

3.4.3 LOINC

LOINC provides a set of universal names and ID codes for identifying laboratory and clinical test results in hospitals[89]. Different hospital uses different standards to capture the laboratory data. When the patient moves from one hospital to another hospital, there are some chance that hospital might ask the patient to repeat the test. During this time, the patient has to repeat the laboratory procedure. In order to avoid the repetition all hospitals' medical procedure should be recorded using LOINC codes as shown in Figure 3.6.

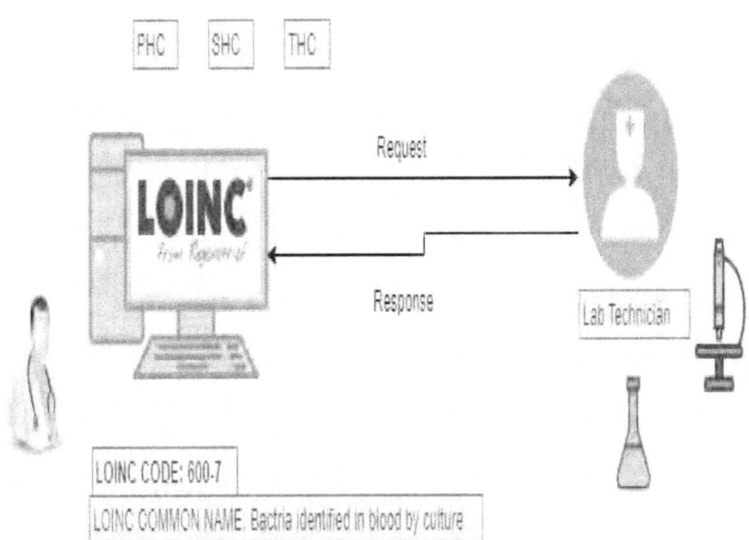

Figure 3.6: Use of LOINC codes in lab test

Recently LOINC team updated the LOINC database by creating 365 new LOINC codes for SARS Coronavirus 2(SARS-CoV-2) testing for the diagnosis of Coronavirus Disease 2019 (COVID-19). Using this code improves the monitoring and tracking of the disease. SARS-CoV-2 is the name of the virus that causes the COVID-19 disease.

3.4.4 DICOM

DICOM is a universal standard used for handling digital medical image in all disciplines of the healthcare facility. The main goal of DICOM Standards is to improve workflow efficiency between the imaging system and another information system. DICOM helps in connecting different health centers together to share information in a standard way by using digital image information. Currently DICOM works as an upper layer protocol that is used in TCP/IP protocol suite[90]. Application layer protocol is independent of the physical network, and with its unique properties enables to communicate any two compatible set of services and information object. Algorithm 1 for viewing DICOM image within EHR system. The DICOM view option module in the EHR system provides healthcare professionals to view and analyze DICOM images through secure web based application. The editing features in the system allows healthcare professionals to adjust the image brightness, contrast etc. for the better view, which results in accurate diagnosis and treatment to the patient.

Algorithm 1 DICOM Viewing Option

```
Begin
file = FS.get(ObjectId(_id))
ds = pydicom.dcmread(file)
shape = ds.pixel_array.shape
image_2d = ds.pixel array.astype(float)
image_2d_scaled = (np.maximum(image_2d,0) / image_2d.max()) * 255.0
image_2d_scaled = np.uint8(image_2d_scaled)
fname=str(file.filename)+ +str(session[num])+.png
session[fname] = fname
with open(fname, wb) as png_file:
w = png.Writer(shape[1], shape[0], greyscale=True)
w.write(png_file, image_2d scaled)
dst = /home/SAHealthcare/healthcare/static/images/
import os.path
if os.path.exists(dst + fname):
os.unlink(dst + fname)
move(fname, dst)
dir_list = ds.dir()
new_dir = fg
for d in dir_list:
if d!=PixelData:
new_dir[d] = ds.get(d,-)
return dir=new_dir
End
```

3.4.5 NDC

NDC stands for National Drug Code(NDC) generally used in pharmacy systems[91]. It is a standardized drug code, mainly used to deliver prescription record to healthcare providers in computer processable format. It is a 10 digit code, consisting of 3 part segments, separated by dashes. The part segments includes labeler, product and package which has the different representations.

- **Labeler Segment(LS):** It is assigned by Food and Drug Administration (FDA) to the manufacturer, distributor of the product.

- **Product Segment(PS):** It can identify the specific drug product. (e.g., Aspirin 325 mg oral tablets)
- **Package Segment(PS):** It is manly used to distinguish packages of different sizes (25 tablets, 100 tables).

In 2001, National library of medicine developed a clinical drug vocabulary called RxNorm in consultation with HL7 technical committee and Veterans Administration. It was developed to improve the standardization in the medication terminology to improve the interoperability by enabling information exchange between different healthcare service providers.

3.5 Patient Work Flow Pattern in Proposed EHR System

Figure 3.7 represents the patient work flow in healthcare levels. When a patient arrives at the hospital at the registration desk, patient demographic details are captured and authenticated using a valid ID issued from the national agency, most preferably, Aadhar Card. Once the registration process is complete, the patient will receive an SMS notification to the mobile. The notification contains patient user name and password. Once registered, the patient moves to the nurse, where the basic vital signs such as height, weight, blood pressure, etc. are collected. On completion, the patient meets the doctor where the doctor takes the history and physically examines the patient and if needed, suggests for the diagnostic tests such as; Lab, Radiology, or others, depending upon the initial assessment, and records the entire event in the EHR database. The investigation result is stored using LOINC, DICOM, and recorded in the cloud-based EHR database. The patient data stored in the EHR database is made available to the doctor for further verification and treatment. Once the doctor confirms the diagnosis and decides on treatment, he/she enters the treatment details into the EHR database using ICD or SNOMED-CT. The doctor then writes a prescription and transfers the details to the pharmacist for dispensing of medicine. The pharmacist verifies the prescription details and records the drug detail using the National Drug

Code (NDC) Database for further communication. During the follow-up visit, the doctors can retrieve the previous treatment history or medication for patient care.

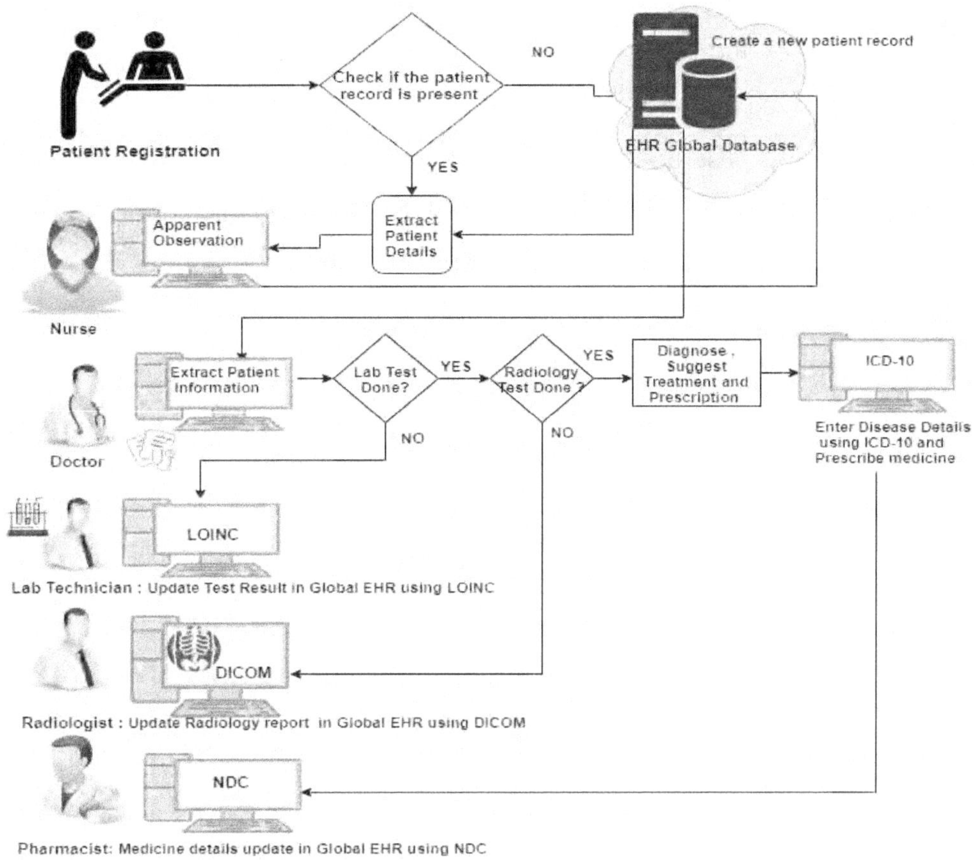

Figure 3.7: Patient work flow pattern in proposed EHR system

Current healthcare information systems of the hospitals are usually isolated from each other as most of the hospitals and healthcare institutions have their own format to create EMR to serve the purpose of treating the patient. But this approach does not support when the patient moves from one health care facility to another for care and consultation. The EMR helps healthcare professionals to use the patient record within the organization. This limitation calls for an Electronic Health Record (EHR) which is built on standard coding terminology and cloud enabled so that the health information generated at various location within and outside the institution can be captured without any difficulty.

Figure 3.8 shows the cloud-enabled architecture for standard EHR. This architecture connects all the users of health information system into one single point to record and share the patient health information. The primary purpose of EHR is to enable efficient and continuous care of the patient. It is a cloud based framework which enables all levels of healthcare professionals to contribute to the EHR contents. In this, cloud resources are shared with different healthcare sectors such as primary, secondary and tertiary. Cloud enabled architecture is used to record patient data in a organized manner which is distributed across different locations. Each cloud serves its purpose of storing the patient record. All the data captured from each healthcare setting is finally stored in global cloud database called EHR universal healthcare service provider database.

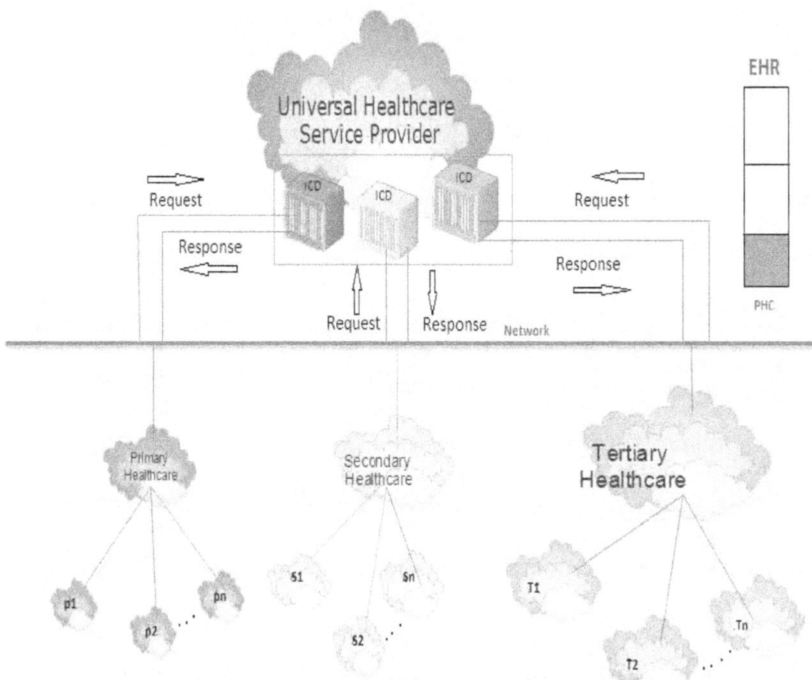

Figure 3.8: Cloud-enabled architecture for standard EHR

The EHR is designed using standard medical terminology and coding standards such as SNOMED-CT, ICD-10, LOINC, DICOM and NDC. Whenever patient visits a hospital including PHC, SHC, and THC, health related data about the patient is generated. The health information of the patient is recorded from a various department, clinical, laboratory, radiology, and pharmacy section for administrative and management purpose. During the diagnosis process, the doctors retrieve the health information of patients and analyze it to diagnose the illness. The system allows healthcare professionals to generate various statistical reports. These reports are presented in terms of the graph and assisted the professional to have a better view and in taking quality decision and appropriate action. Doctors from primary health care can take expert advice from secondary or tertiary care specialists by sharing the information. Current model makes use of RBAC in order to access data from different levels of the healthcare system. Using this approach, the three levels are given access criteria such that primary level care doctor can see patient data of primary hospital, secondary level care doctor can see both primary and secondary hospital data of the patient visited and finally tertiary level care can see data (write/update with permission) from all hospitals connected to EHR database and it is also applicable to other way round from tertiary to primary care with access permission.

3.6 EHR System Design

In software development life cycle, design is the most essential phase. System designer needs to carryout careful planning and thinking which helps to understand how various parts of the system are going to achieve the desired goal. Design should be done with extreme care because during this stage if it has any errors, then that will affect the overall performance of the system. During requirement analysis phase, requirements are collected and specified in the document. Design is the second phase where it can provide a solution to the problem. This phase contains set of iterative steps which enables the designer to define all features of the system to be built. Design basically acts as a bridge between system requirement specification and final solution for satisfying the requirement[92]. The design of a system is essentially a blue print or a plan for the solution of the problem by the system. The content required for developing EHR system is classified into various sections meant to capture vital aspect of a patient's health and medical status. The main module incorporated in the EHR system is the database design depicted in Figure 3.9.

Figure 3.9: Database Design Diagram

3.6.1 Patient demographics

One of the main responsibility of the healthcare professionals is to collect the demographic details of the patient accurately when a patient visits the hospital for the first time. Collecting demographic details in consistent manner help the hospital staff to address the unique needs of the patient. The patient demographic section includes details such as name, address, email, contact number, gender, marital status, occupation, emergency contact, location, previous diagnosis etc. The most important part would be a unique identification of the patient. EHR also has a support of including medical insurance and healthcare financing details.

3.6.2 Patient health history and medications

The EHR system allows healthcare professionals to capture the patient's past medical history and present symptoms. Details of past illness such as hypertension, diabetics, kidney disease, heart disease, asthma, tuberculosis and other needs to be recoded for better diagnosis of the current illness. The various lab tests and multimedia images need to be present to the doctor along with the EHR contents. The drug history mainly captures the list of drugs that the patient was taking prior to the admission in the hospital. In addition to this, EHR system has provision to provide details of drug allergies, dosage, duration, recently stopped medicine. Addiction history such as smoking, tobacco, and alcohol details need to be recorded. The health history of the patient and the patient's family helps to address acute health problems and minimize the chronic condition of the patient.

3.6.3 Patient clinical encounters

Clinical encounter provides a principal focus to provide patient centric medical care. During clinical encounter healthcare professionals can get quality data to treat the patient better. Details for the clinical encounter are captured in different healthcare levels and the details are displayed when patient visits the hospital for the current treatment. Details includes physical examination, observations, diagnosis, lab investigation, encounter(interaction between the patient and doctors) type, name, time need to be captured.

3.6.4 Legal and regulatory compliance

Patient and healthcare professionals are the primary owners of the EHR. It is a confidential communication between patient and healthcare staff and it cannot be released without the permission. A health record is the most important legal and business record for healthcare organization. All reports i.e. lab records, X-ray , ultrasound , magnetic imaging resonance (MRI), Computed Tomography (CT-scan), and prescription details should be signed by a doctor before issuing the document or transferring the record from one healthcare levels to others.

3.7 EHR Product Features

3.7.1 General description

Cloud based EHR system is a web based application to create and manage the patient health information at different levels of healthcare system. The system should have the features to make the data available to all the end users and generate the statistical report of the patient visited to PHC, SHC, THCs.

3.7.2 Purpose

The purpose of developing this type of product is to enable patient centric approach where distributed patient health records are made available in an integrated and unified manner at the point of care.

3.7.3 Scope

Cloud based EHR web application is developed to manage patient health information of the outpatient department or clinic. Registered users must be able to login their accounts, after approved by the administrator, accessing the web application and signing with their user name and password anytime.

3.7.4 User characteristics

The target users are administrators, healthcare professionals, and patients from primary, secondary and tertiary healthcare facilities. Two factor authentication is incorporated for each user to provide higher level of security. This prevents unauthorized access to the EHR system. The administrator is responsible for monitoring and managing the database activity of healthcare users. The patients need a training to use the web forms to perform different activities.

3.7.5 Specific Requirement

Careful designing is done on methods and tools to be used. The developed model uses MongoDB Atlas Cloud Cluster. All of the data boils down to multiple collections which are stored in a cluster in the cloud. For retrieval operations, information is extracted from the specific collections in the cluster, then processed using algorithms and then displayed to the end-user. For storing, data which is generated from user inputs, is retrieved from the web application and processed to attain certain standard formats before storing in the collections. In this way a systematic data flow is established and sanctity of the model is maintained. Web based EHR system that is developed is platform independent application which runs on any computer, laptop, mobile phone which is connected to the Internet.

3.7.6 Security requirement

Cloud based EHR web application allow only registered and approved users to use the functionalities of system. EHR has the provision to manage the privacy of the healthcare users. To access healthcare services, user needs to enter valid user credentials during application access. Details of the security requirement adopted in the developed system is discussed in detail in Chapter 4.

3.7.7 Constraints

To access the healthcare service requires fully functional Internet connectivity with 24-hour server uptime for optimal functionality. In case, if the end user loses the Internet connectivity, the error message like "Temporary lost connection" will be displayed. The end user has to perform the task again on the resumption of Internet connection.

3.8 Use Case Specification of EHR system

The Use case is a methodology to identify, clarify and organize the system requirement. The main expected functionalities of EHR system are shown in the form of use case diagram as shown in Figure 3.10. The purpose of the use case diagram is to show the interaction between the actors and the system. The EHR consists of major actors such as the administrator, patient, receptionist, nurse, doctor, lab technician, radiologist and pharmacist. The EHR system is built using Role Based Access Control where the system allows the user to access service only with role assigned by the system, so not all healthcare users can access the entire services of the system.

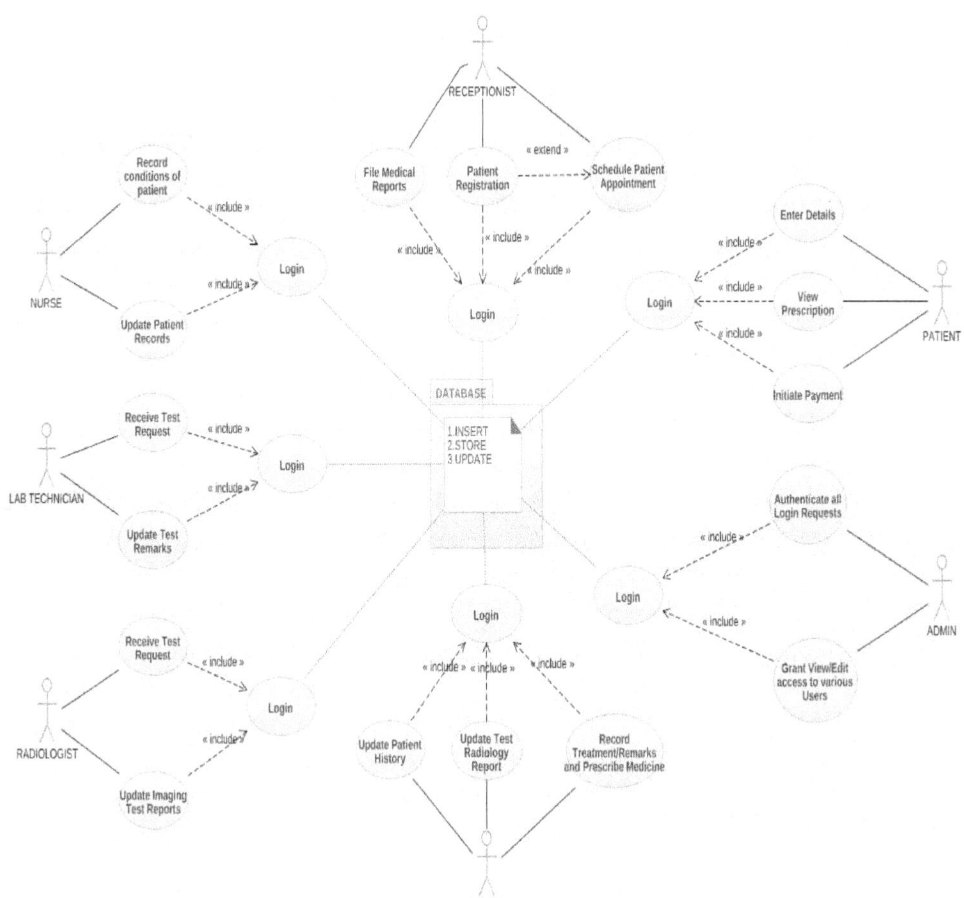

Figure 3.10: Use-case diagram

The use case specification is a very important step in EHR design. There is a necessity to understand the critical needs of the users and users functionalities to develop a standardized complete view of the EHR system. Use case specification captures the specific details of the use case. It provides a way to capture the different scenarios that exist in the system. Use case specification gives the detailed methodological description which is not provided with the help of use-case diagram.

Table 3.7: Use-case specification for login

Use Case 1	Login
Scope	Web application
Primary Actor	All users of healthcare ecosystem
Stakeholders and Interests	Hospital Administrator Receptionist Patient Nurse Doctor Lab Technician Radiologist Pharmacist
Preconditions	Access the web application using URL and application is loaded into the cloud based system
Main Flows	1. The use case begins when healthcare users enter the URL in search engine and web application loads from cloud system. 2. The home page of the web application allows (text box) user to enter user name and password, including submit, cancel option. 3. The user needs to enter the login credentials. 4. Once the user clicks the submit button. 4.1 If the submitted login credentials are invalid a.The user needs to enter login credentials again b. Checks the login details for validation. 5. The web application allows users to perform the required operation.
Post conditions	The Application is available for healthcare users to record health data.
Alternate Flow	1. If the user is unable to perform the required option. 1.1 Then the user will be prompted again to try the login (maximum three times allowed). User returns to step 2

The various use case specifications of different healthcare users and their functionalities in the EHR system is depicted in Table 3.7, 3.8, 3.9, 3.10.

Table 3.8: Use-case specification for Patient registration

Use-case 2	Patient Registration
Scope	Web application
Primary Actor	Administrator/Receptionist
Stakeholders and Interests	Hospital Administrator Receptionist Patient Nurse Doctor Lab Technician Radiologist Pharmacist
Preconditions	The person should be registered and must have a valid login credentials in the web-based EHR application.
Main Flows:	1. The use-case begins when hospital admin/receptionist selects the option in main menu for patient registration. 2. Web application provides the user with a required field by displaying the form. 3. When the entered details are invalid. 3.1 Web application displays error messages to enter details again 4. The application prompt entry fields again and validates the all the fields. 5. Once all the fields are validated new patient details are recorded in the EHR database.
Post conditions:	Application populates the form with the patient details
Alternate Flow:	1. If the user is unable to perform the required option. 1.1 Then the user will be prompted again to enter the valid details. User returns to step 2

Table 3.9: Use-case specification for managing patient record

Use-case 3	Managing patient health information in EHR database
Scope	Web application
Primary Actor	Receptionist/Nurse/Doctor/Lab Technician/Pharmacist
Stakeholders and Interests	Hospital Administrator Receptionist Patient Nurse Doctor Lab Technician Radiologist Pharmacist
Preconditions	The person should have a valid login credentials and the patient record should be created.
Major Flows	1. The use-case begins when a patient visits healthcare facility for the treatment. 2. Receptionist confirms the doctor's availability and gives an appointment to the patient. 3. The receptionist transfers patient record to in-charge nurse for recording vital signs and general checkup. 5. Nurse transfers patients' records and observations to the doctor 6. Nurse call the patient to consult the doctor at the doctors' office 7. The doctor examines the patient and views the previous health history of the patient. 8. Doctor orders lab tests using LOINC code. 10. The lab technician receives the LOINC code and performs the lab test and updates the patient record. 11. The doctor views the lab record and writes the diagnosis for the patient using ICD10/SNOMED CT code. 12. The doctor creates medication detail using the NDC code.
Post conditions	Patient health record is updated with all current visit details
Alternate Flow	1.If the healthcare user is unable to perform the required operation. 1.1 Then the healthcare user will be prompted again to enter the Lab test/Diagnosis/Medication code details. User returns to step 8 for the lab test, step 11 for diagnosis code and step 12 for entering medication details.

Table 3.10: Use-case specification for Managing Prescription

Use-case 4	Create Prescription
Scope	Web application
Primary Actor	Doctor/Nurse/Pharmacist
Stakeholders and Interests	Hospital Administrator Receptionist Patient Nurse Doctor Lab Technician Radiologist Pharmacist
Preconditions	The person should have a valid login credentials and patient diagnosis and treatment history needs to be retrieved.
Main Flows:	1. The use-case begins when the user clicks create medication 2. Web application makes provision to enter medication details using NDC. 3. When the user selects the medication name 3.1 The application displays the related NDC code for the medicine 3.2 The user selects the dose and other details 4. Once all the fields are entered new patient details are recorded in the EHR database. 5. Access is given to the pharmacist to assist the patient to give proper medicine.
Post conditions:	Application populates the form with the medication details
Alternate Flow:	1. If the user is unable to perform the required operation. 1.1 Then the user will be prompted to enter the medication details. User returns to step 2

3.9 Overall Description of EHR System

Cloud based EHR application provides web based forms to capture patient health information. These forms will be processed and served as the history of the patients treatment over a period of time. Cloud storage helps to attain a global platform which is accessible from every part of the world independent of time and date. These few factors guide the model in attaining EHR fluidity by making patient records globally available to verified users.

Figure 3.11 depicts overall activity workflow of the EHR model. The main idea behind developing the cloud based is to ease the transfer of EHRs so that patients are exempted from the burden to carry their medical record with them whenever they change the hospital.

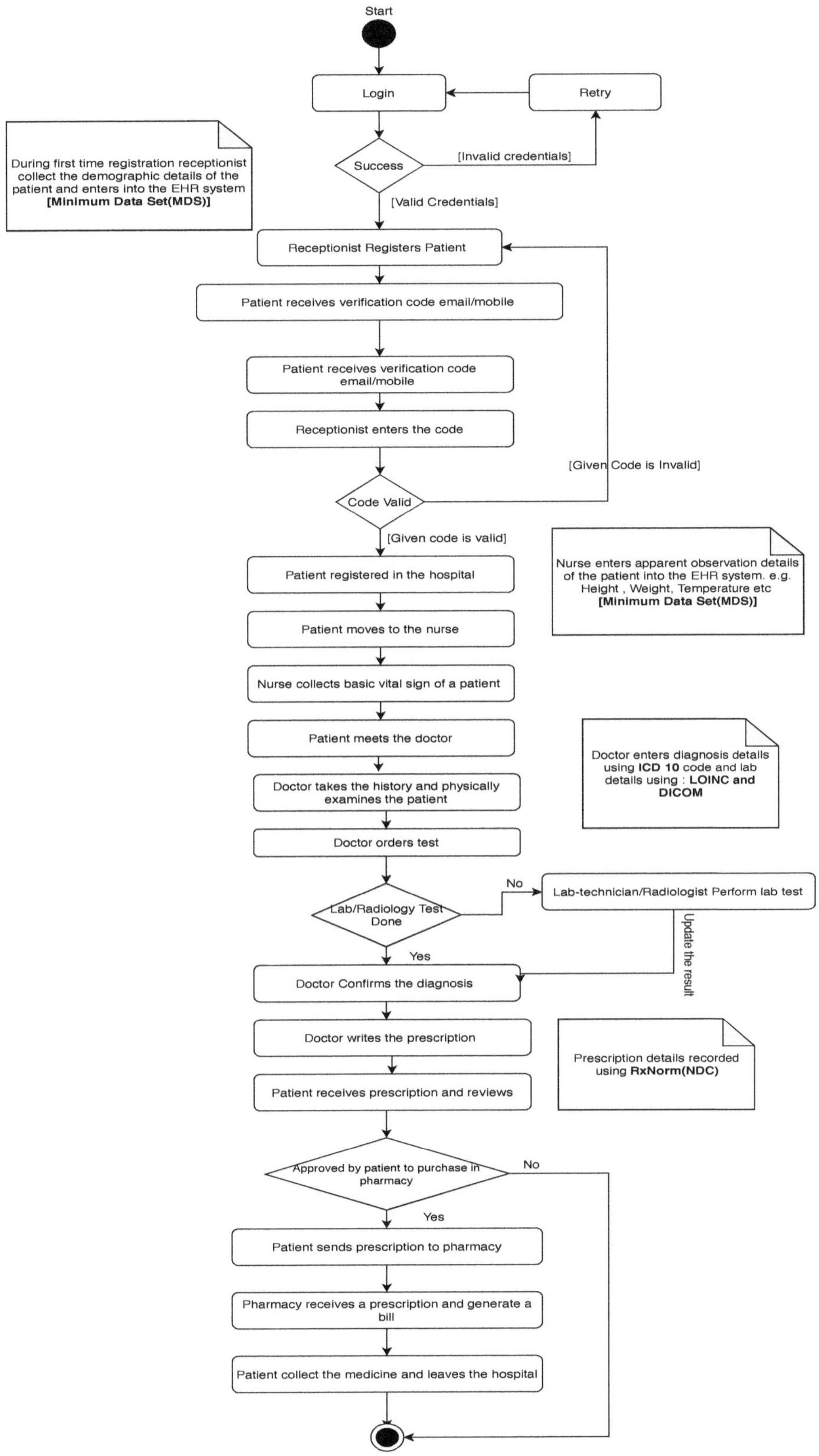

Figure 3.11: Overall activity work-flow of EHR System

Electronic record management and record transfer is achieved through a web application and cloud based data storage. The following are the list of healthcare users and their activity in the developed EHR.

- Patient is registered on the web site by the administrator and assigned to doctor.
- Nurse takes basic health history of the patient
- Doctor diagnoses(ICD-10 and SNOMED-CT codes) the patient and orders test, medication(NDC)
- Lab technician/Radiologist performs lab test and updates the report on the website with LOINC Codes,DICOM.
- Pharmacist receives the order and pharmacy gets the order ready.

The login activity performed by the healthcare users are shown in the Figure 3.12.

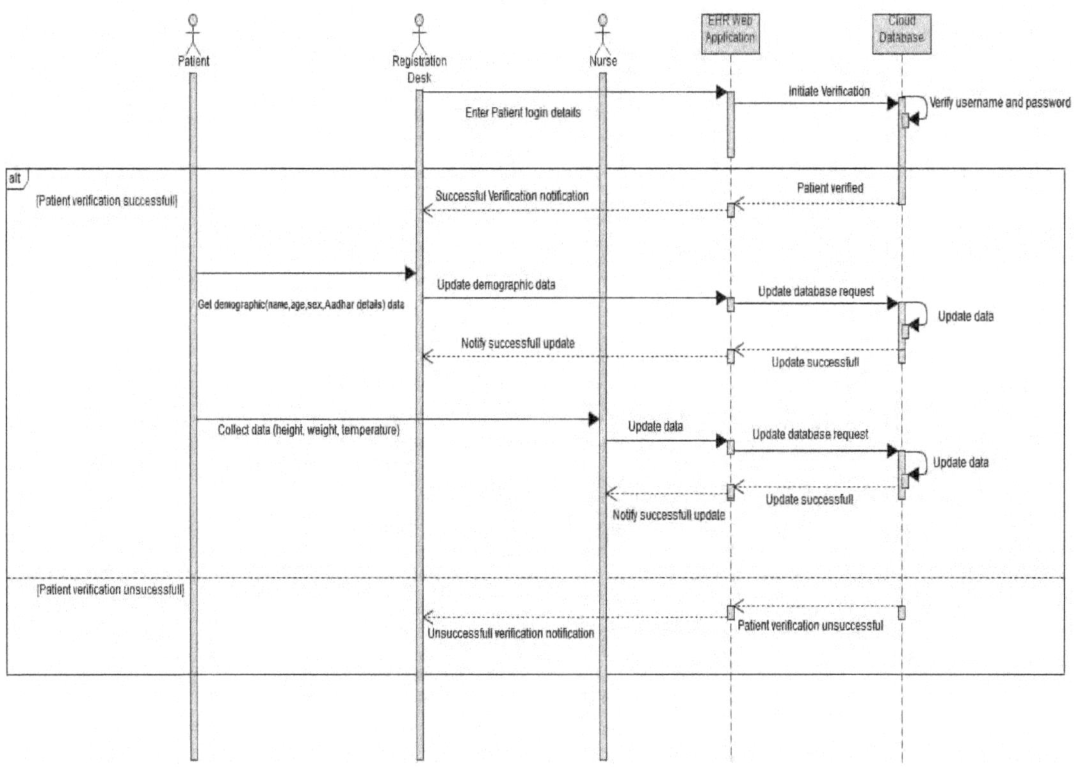

Figure 3.12: Sequence of activity during patient registration

During first time hospital visit, a patient health record needs to be created in EHR web application. The registration staff collects the required information and creates the patient record in the EHR system. During registration the unique patient id is generated with password. Authentication of the patient is done at the cloud server and the patient data is stored in encrypted format. Each health-care user can request for patient record depending on the how much access features are enabled for each user.

The sequence of activity performed by the nurse in EHR system is shown in the Figure 3.13. Once the registration process is complete, the patient moves to the nurse, where the basic vital signs height, weight, blood pressure, etc are collected. After recording the patient details in the EHR system , the sensitive details are encrypted and stored in the cloud database. On completion, the patient is sent to the doctor.

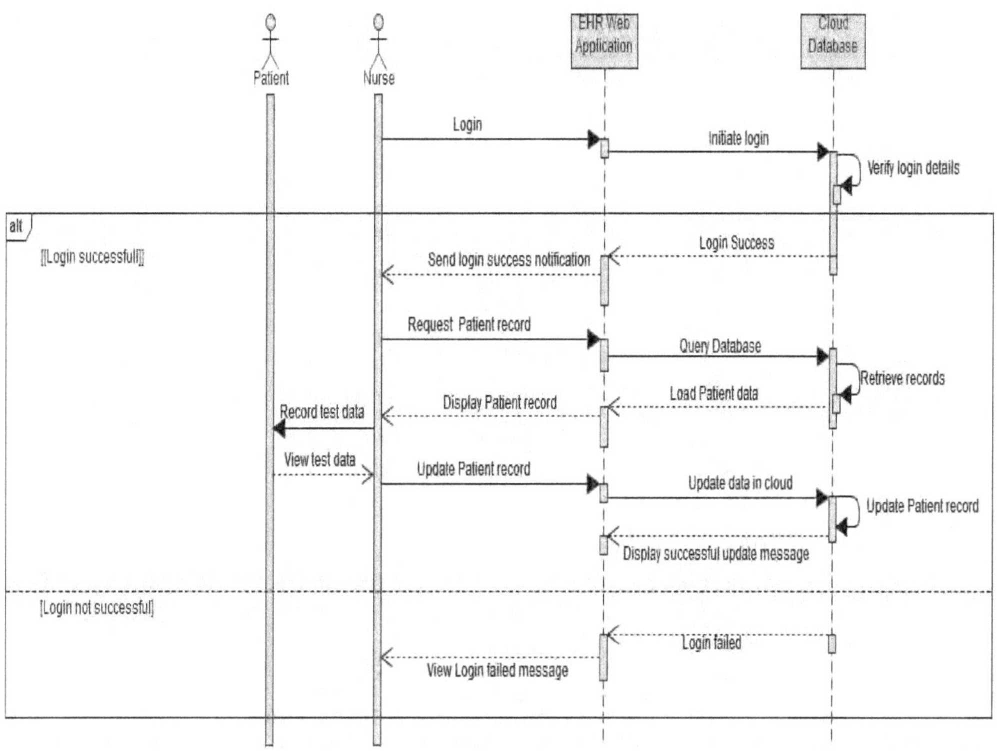

Figure 3.13: Sequence activity of nurse

The sequence of activity performed by the doctor in EHR system is shown in the Figure 3.14. The doctor takes the history and physically examines the patient and suggests for the diagnostic test such as Lab, Radiology or others depending upon the initial assessment, and records the entire event in the EHR database. Data generated from the laboratory test is stored using LOINC, DICOM and recorded in the cloud-based EHR database.

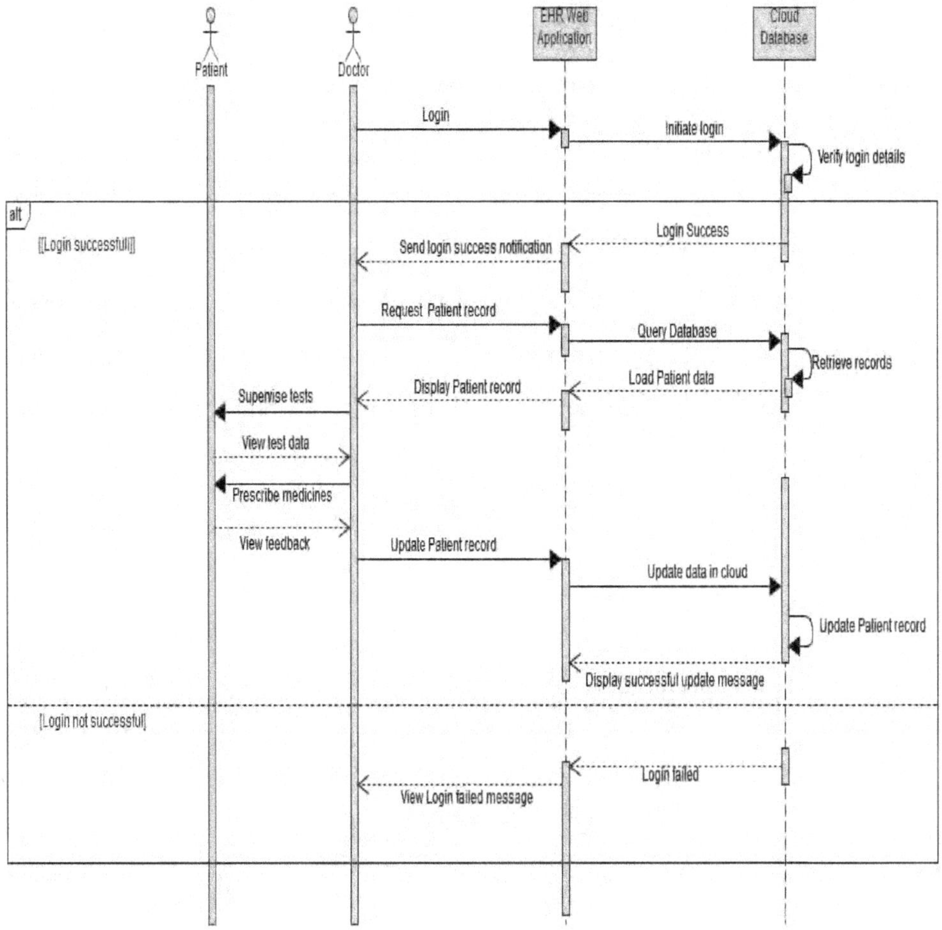

Figure 3.14: Sequence of activity of Doctors

The sequence of activity performed by the lab technician in EHR system is shown in the Figure 3.15. The patient data stored in the EHR database is made available to the doctor for further verification and care. Doctor confirms the diagnosis and enters the treatment details and symptoms into the EHR database using one of the classification standards like ICD or SNOMED-CT. The doctor then writes a prescription, and that data is transferred to the pharmacist for dispensing of medicine.

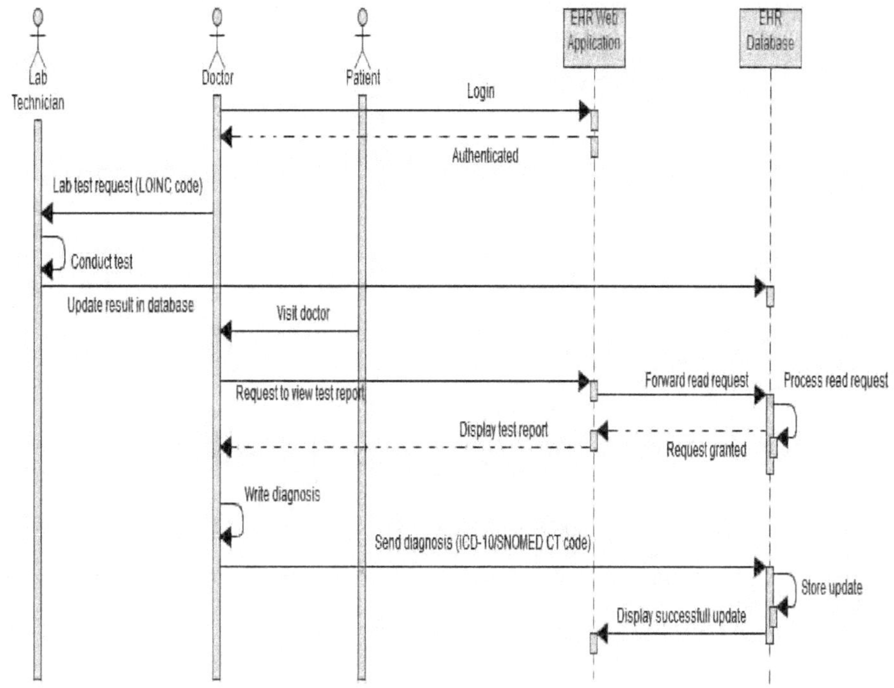

Figure 3.15: Sequence of activity of Lab Technician/Radiologist

The sequence of activity performed by the pharmacist in EHR system is shown in the Figure 3.16. The main responsibility of pharmacy system is to manage prescription written by the doctor and dispensing of medicines. The pharmacy module in the EHR system uses NDC standard coding database to record and report the drug details.

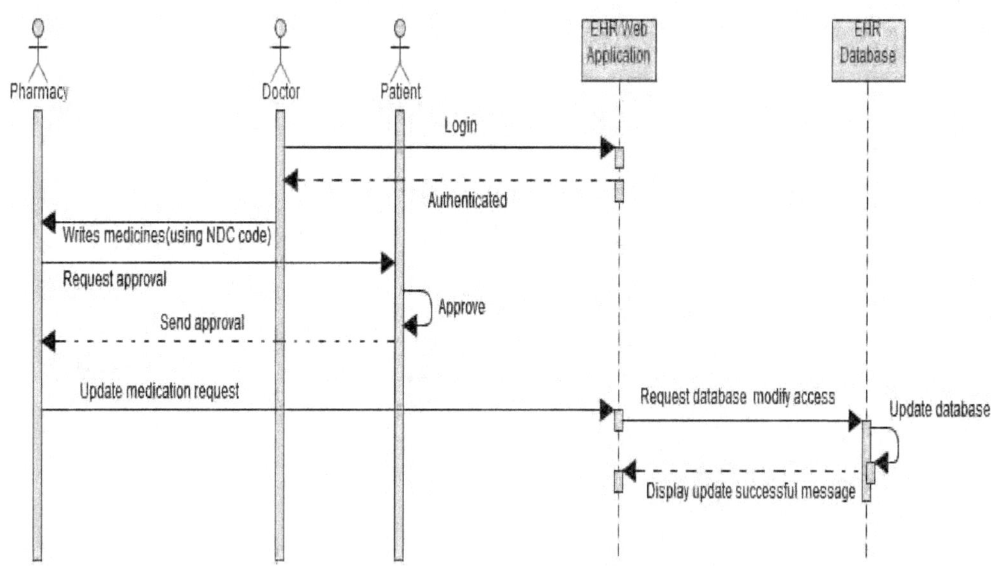

Figure 3.16: Sequence of activity of Pharmacist

Doctor writes the prescription to the patient using NDC. Same code is received by the pharmacist. The provision is given for the the patient to purchase the requested medication in pharmacy or outside and an approval request is sent to the patient. If the patient approves the request then medicines will be given to the patient by the pharmacy. The pharmacy module has the complete history of the patient related to medication including drug name, dosage, start data, stop date, prescribed by etc.

The developed and deployed system provides the following features to the healthcare users:

- Every healthcare user with proper authentication is allowed to store and access the healthcare records. Also, they can keep track of all their medical activities, right from viewing their medical records to viewing prescriptions and ordering medicines, everything can be done in one place.
- EHR application supports health-care standard codes such as SNOMED CT, ICD-10, LOINC, DICOM and National Drug Code(NDC). The radiology image format such as DICOM can be uploaded directly from individual local systems like a normal upload. To operate with DICOM image, doctor has to click a button to view the image and it will display the tag associated with it for detailed diagnosis.
- The process of writing prescriptions and ordering tests has become as easy and efficient for the healthcare providers.
- The developed EHR application has user-friendly UI with few button clicks to perform multiple tasks and also provides a platform where everything pertaining to healthcare can be easily accessible in one centralized location.

3.10 Time Motion Study: Conventional and Developed EHR system

Figure 3.7 discussed earlier in section 3.5 represents the patient work-flow in healthcare facility of PHC, SHC and THC in which easy and immediate access of patient data is evaluated with the help of time motion study. Healthcare standards

Table 3.11: Comparative list of conventional work-flow pattern with proposed EHR system

Process	**Existing**	**With Electronic Health Record (Estimated)**
Registration of patient data	Patient information recorded manually on paper. **10 Min(Avg.)**	Patient information recorded electronically in EHR system and is accessible immediately at any location
Amount of time required for OPD section to receive medical records from MRD	Respective department receives the medical record from messenger boy **20 Min (Avg.)**	Instant access to patient information by the clinicians by passing the unique patient ID into the respective module.
Recording patient vital signs into the medical record	Clinician requires more time for recording patient data manually	Web enabled electronic recording requires less time for recording patient vital sign.
Recording lab investigation result report of the patient	Manual entry needs messenger boy to handover the report to lab technician **20 Min. (Avg)**	Instant access to investigation results at the clinician system as and when it is entered into the system.
Recording patient diagnosis details into the medical record	Clinician(Doctor) requires more time for recording patient data manually	Web enabled electronic recording requires less time for recording patient disease details.
Access of recent clinical knowledge	Difficult to search required information on manual basis.	Immediate access of investigation result at the clinician system as and when it is entered into the system.
Transfer of record from one healthcare levels to other	The patient should carry hard-copy of the medical record with an authorized signature.	If the healthcare professional at the referral level wish to see the patient complete record, an instant transfer request is sent to the level from where the patient is referred and upon receiving the authorization, he/she will have access to patient data.

and terminology play a major role in seamless sharing of data between many levels of healthcare facility. The following interpretations were drawn by visiting the primary, secondary and tertiary care hospital. The analysis (Table 3.11) of the work-flow of the patient and feedback received from the health care professionals discloses that the traditional system requires more time to access patient data during the clinical encounters. Manual based entry takes maximum time in storing and retrieval of patient records from Medical Record Department(MRD) to Outpatient Department(OPD). This results in increase in waiting time of the patient who needs immediate care. Accessing the previous history of the patient and ordering laboratory test for the patient all are manual and it increases clinician time in documentation. In contrast it is felt that adopting Electronic Health Record in Indian health-care scenario will positively impact the information service support in various levels of healthcare system.

Summary

Detailed methodology adopted for the development of standardized EHR in all levels of healthcare systems are discussed. The healthcare facility visit helped to understated the limitation of the system. The MDS and healthcare standards adopted in cloud based EHR system has a provision to share data(interoperability) among various levels of healthcare system. The subsequent chapter discusses the detailed security framework adopted for securing cloud based EHR.

Chapter 4

Security Framework For Cloud Based EHR System

4.1 Introduction

Advancement in IT technology in healthcare sector plays an important role to maintain and manage the patient health data in digital form in any levels of healthcare system. Cloud Computing (CC) plays a major role in usage of wide range of services including interoperability and standardization in healthcare ecosystem. Every time a patient visits healthcare facility, health information of the patient is recorded and stored in the cloud data repository. As a result, the mass storage of sensitive information about the patient in the cloud storage is considered as one of the most targeted sector by the intruder. It is very important to take special care in terms of security and privacy of the EHR system. Storing EHR data in the cloud can be quite dangerous if not properly protected and can lead to leaking of sensitive information, and also affect the doctor's diagnosis. Protection of health information should be a primary concern during the development and implementation of any IT application in a hospital setup. The issues related to security and privacy of EHR system is identified, and measures are taken during the design, development, and implementation of EHR application. The rest of the chapter is organized into different sections as shown in Figure 4.1.

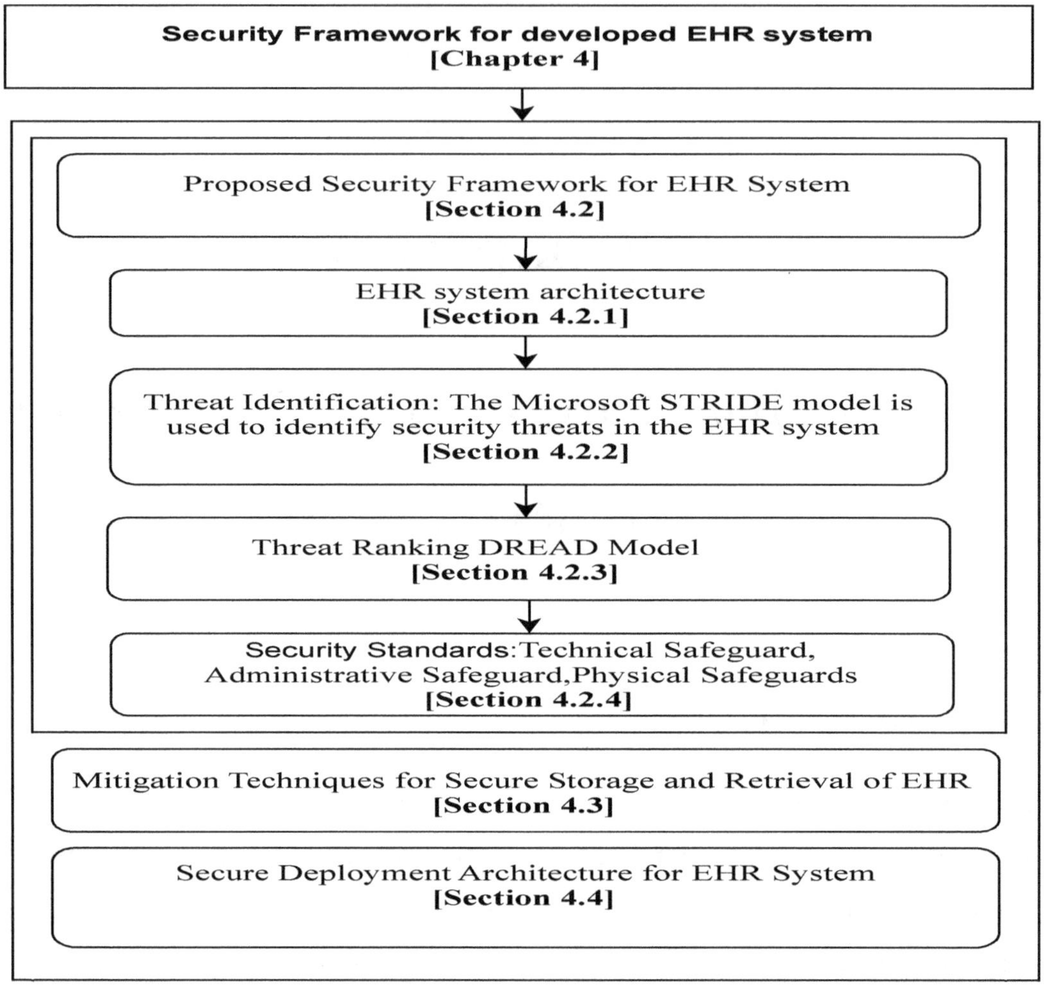

Figure 4.1: Organization of chapters into different sections

4.2 Proposed Security framework for EHR System

Healthcare organizations are considered as one of the most targeted sectors by intruders to gain sensitive information. To share the data among multiple healthcare providers in a secure manner, EHR reference architecture is developed and various attacks and vulnerabilities have been identified. The security framework for EHR system is based on threat modeling and is depicted in Figure 4.2. Different components of the security framework for an EHR system are discussed in the following sections

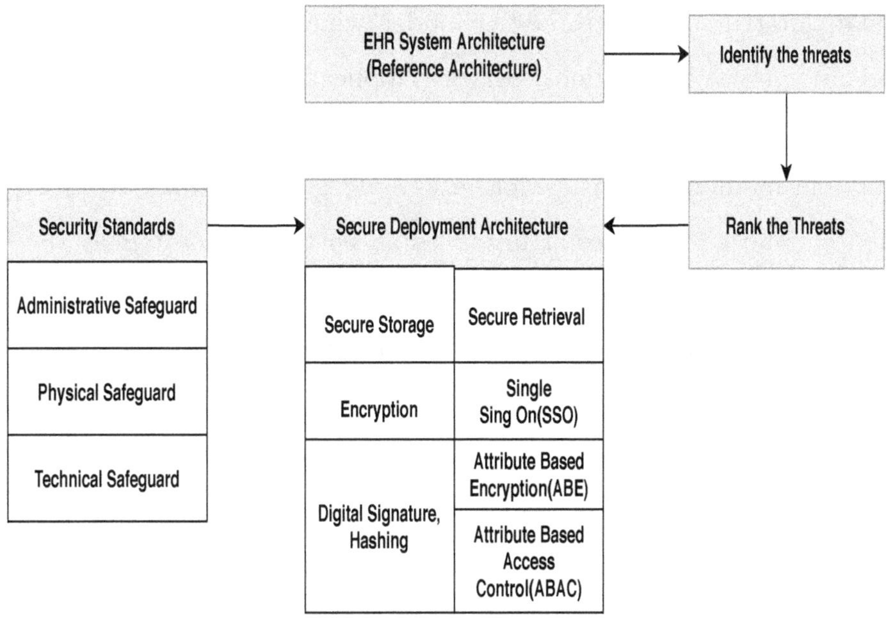

Figure 4.2: Proposed Security Framework for EHR system

4.2.1 EHR system architecture

EHR has the potential to provide real-time data of the patients by connecting various levels of the public healthcare facility, irrespective of their geographical boundaries. The security of the EHR is the major concern when sharing health information at various healthcare levels. The reference architecture for identifying threats in the system is depicted in Figure 4.3.

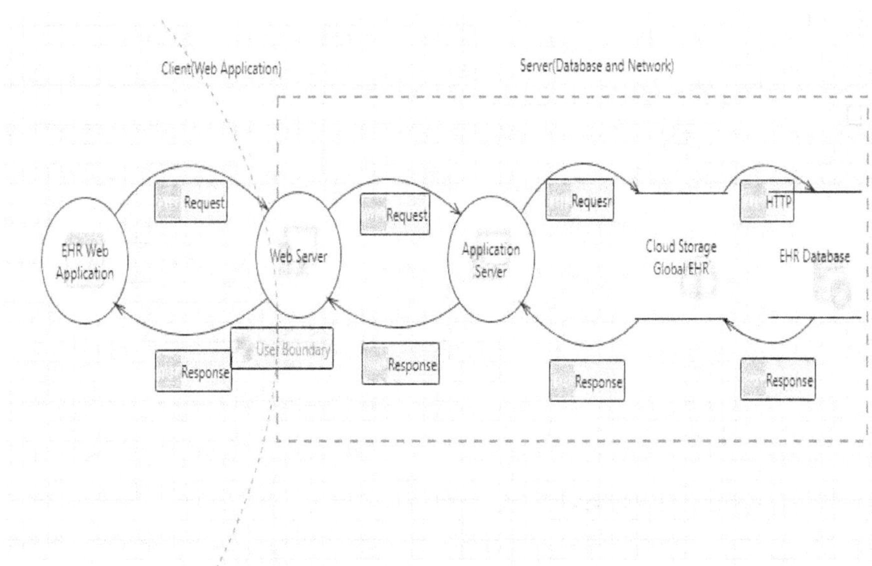

Figure 4.3: EHR system architecture

The architecture follows three-tier model, which is divided into three major parts such as client (Web application), Server (Applications and Database), and networks. The first visible part of an EHR system is the client. It plays a key role for viewing, entering and modifying health information. This component is thus expected to be attacked the most. The second most visible part of the system is the server. The compromising server gives complete control to the attacker. The server attacks include revealing, changing and destroying health information which completely harms the system. The third component is the network. A compromise in the network results in eavesdropping, altering and more when data is in transit mode. The attacker's main goal is to target the visible component of the system including client, server, and networks. Compromising any one of them result in compromising the entire system.

4.2.2 Identifying Threats

The Threat Modeling(TM) Tool [93] is used by software architects to determine and mitigate possible security threats in the system. Microsoft Security Development Life Cycle (SDLC) is a core component in development life cycle which provides cost-effective way of developing a system with the minimum number of security errors. TM is the process that improves application and network security by identifying and rating the potential threats and vulnerabilities in the EHR system. TM is especially important during the design phase. TM allows experts to determine the possible threats that could affect the system. The reference architecture designed for EHR system is given as input to the threat model to identify threats which helps to develop meaningful security model for cloud-based health information system. Threats exist due to weaknesses in design, implementation or configuration. These threats are identified by a systematic review of the system which helps in finding the intruder access points in order to eliminate the threats. The Microsoft STRIDE model is used to identify security threats in the proposed EHR system. The possible threat category list for the health information system is shown in Table 4.1.

4.2.3 Threat Ranking

The identified threats are ranked using DREAD. The DREAD is an acronym built from the first letter of each category of threat namely Damage Potential(D),

Table 4.1: STRIDE model threat category list for the EHR system

Sl.No	Threat Category	Description
1	Spoofing(S)	With other user identities, the attacker successfully impersonates the account in the EHR system
2	Tampering(T)	Performing improper alteration in the EHR record
3	Repudiation(R)	Attacker has the skill set to hide their malicious activity
4	Information Disclosure(I)	Sensitive patient information is exposed
5	Denial Of Service(D)	The required resources are inaccessible to the intended users of the EHR system.
6	Elevation Of Privilege (E)	Attacker gets additional permission to access the content of EHR system

Reproducibility(R), Exploitability(E), Affected users(A),and Discoverability(D). The risks are rated as DREAD structure with numerical values of 3, 2, and 1. The threat rating scheme is shown in Table 4.2.

Table 4.2: Threat Rating Scheme

Threat Rating	Description
Damage Potential(D)	1 = Nothing 2 = Individual Users Data 3 = Complete System
Reproducibility(R)	1 = The attack is very difficult to reproduce. 2 = Assigning certain time intervals 3 = The attack can be reproduced very easily
Exploitability(E)	1 = Attacker need in-depth knowledge about the system 2 = A skilled programmer can make an attack 3 = A person who is new to the system can exploit
Affected Users(A)	1 = Very small percentage of the users 2 = Some users 3 = All users
Discoverability(D)	1 = Required controls do not exist 2 = Insufficient log management 3 = Full control over system

When threat is high, it means it needs to be monitored continuously and resolved by implementing appropriate security measures. If the identified risk is medium, it also needs to be addressed. If identified risk is low, it does not pose any significant damage to the system.

Equation used to compute a risk value, using DREAD is an average of all five threat categories as shown in equation (1)

$$Risk = (D + R + E + A + D)/5 \tag{4.1}$$

The computed Risk value is categorized as low (0 to 6), medium (7 to 11) or high (12 to 15) which gives an impact the threat poses to the EHR system. DREAD uses the standard scale for computing High, Medium and Low to rate the threat. When threat is high, it means it needs to be resolved by implementing proper security measures. If the identified risk is medium and low it also needs to be addressed. TM is an important part of the application life cycle especially during the design part. The reference architecture is the base model to identify the threats and all threat information is automatically generated by the Microsoft tool. The system generates more than 100 threats for the given reference architecture and few of them are used for computing the risk as shown in Table 4.3 and in Figure 4.4. TM will allow experts to determine what could be the possible threat to the system. Identifying threats helps to develop meaningful security model for cloud based EHR.

Table 4.3: DREAD RATING

Threat(T)	D	R	E	A	D	Total	Rating
T1: Attacker gets authentication credentials by observing the network	3	3	2	2	2	12	HIGH
T2: Theft and replay of authentication cookies	2	2	2	2	2	10	MEDIUM
T3: Links to sites that use cookie less session state	2	1	2	2	1	8	LOW
T4: Possible way of gain access to session IDs	1	1	1	1	1	5	LOW
T5: The attacker obtains sensitive patient data from cloud database	3	3	2	3	2	13	HIGH
T6: Attacker with inadequate authorization is able to see other patient data and may be able to access other restricted data.	3	3	2	2	2	12	HIGH

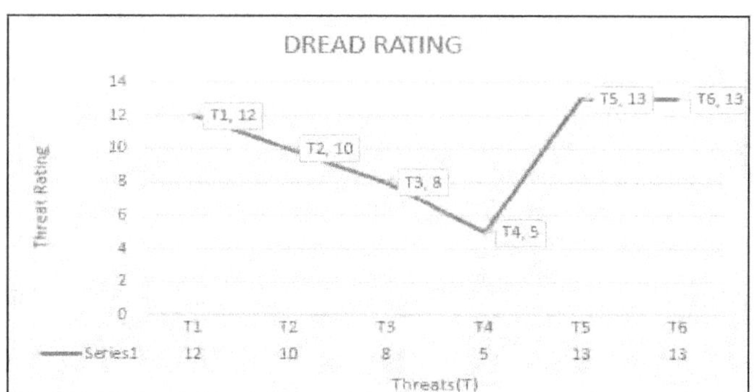

Figure 4.4: DREAD rating with graphical representation

The result indicated the high possible threats related with user authentication and authorization whereas theft and replay of authentication cookies and link to the site of those cookies are found with medium threats. The result also indicated that prohibited users can gain unauthorized access to the health information system by spoofing the login credentials. This threat may lead to the sharing of patient's personal and health-related data in an unauthorized manner that may result in information misuse, information disclosure, and altering information. To enhance security features, Administrative, Physical, and Technical security standards are also considered.

4.2.4 Security standards

The health record system in India necessarily is safeguarded by the privacy and security standards. These standards guarantee that, health data is secure during its storage, retrieval, and transmission. For adopting security standards, the health organizations must consider various aspects such as the size of the organization, complexity of the system, and healthcare IT infrastructure capabilities. Administrative, physical, and technical are the three main safeguards for security standards used in the health information system which ensures confidentiality, integrity, and availability of the EHR system [94]. In this research main focus is given for technical safeguard and the other two is part of the policy and administration related.

4.2.4.1 Technical safeguard

The developed EHR system must comply with technical safeguard to protect its security plan. Technical safeguards includes technology, policy, and procedures for the protection of the EHR by controlling access to it. The proposed system addresses the security functionalities as shown in Table 4.4.

Table 4.4: Technical safeguard for EHR system

S.l No	Type	Purpose
1	User Authentication and authorization	Privilege Management and Role based Access Policy
2	Audit Trail and Logs	All activities performed by the user must be recorded
3	Data Integrity	Secure Hash Algorithm (SHA), recommended using SHA-256 or higher
4	Data Encryption	Encryption key-Minimum 256-bits key length Encrypted connection- HTTPS, SSL v3.0, and TLS v1.2
5	Digital Certificate	Digital certificates use and management

4.2.4.2 Administrative safeguard

Administrative Safeguard mainly safeguards the health information of the patient by implementing proper security process that mainly focuses on policies, and procedures internal to the organization. To ensure security measures and to protect patient information, it is necessary to update the organizational policies. Frequent changes to policies will ensure that intruders get less time to adapt and find possible vulnerabilities. Training the employees to follow security procedures increases their awareness and allows them to take the appropriate actions and follow the established contingency plans. Administrative policies should also clearly mention the access control policies and role to access health data [95].

4.2.4.3 Physical safeguard

To protect the EHR system from physical damage or theft mainly requires physical safeguard technique such as installing closed circuit television (CCTV) cameras. Implementing biometric authentication schemes such as fingerprint login or iris scans can further enhance security. Integrating physical safeguards with cloud based EHR system reduces unauthorized intrusion and limits remote attacks on the network. For instance, if a policy is implemented which allows the generation of a valid login key only after successful biometric authentication, it can narrow down the intruders reach. Physical safeguard also includes the entities which are authorized to grant access to physical resources and their maintenance[95].

4.3 Security mechanism for Secure Storage and Retrieval of EHR

4.3.1 Authentication Model for EHR

Single Sign-On (SSO) is an authentication service for EHR users which allows accessing multiple health services after signing in only once [96]. This avoids multiple logins to access essential health services after successful identification. Figure 4.5 illustrates SAML-SSO based authentication model wherein SSO is implemented using Security Assertion Markup Language (SAML). It is an XML based communication protocol facilitates the sharing of authentication information between the service provider and identity provider (IdP). A small request is generated and redirected to the IdP when any user tries to access the service from the EHR system. IdP contacts active directory present in the hospital and parses the request and authenticates the user for access. During this process, the user is provided a generated token which is used to access the EHR service, thus improving the security and access efficiency.

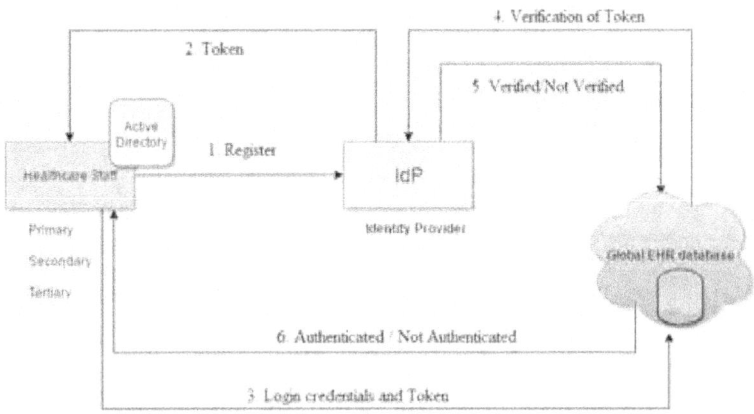

Figure 4.5: Single Sign-On (SSO) Authentication Model for Cloud-Based EHR

In addition to the authentication model, the salting technique which is a random data is used as an additional input to the hash function. Without salting technique, an intruder can precompute the rainbow tables of common password hashes and easily compare them to a database and see who used which common password. With rainbow table attack output of a hash function is always the same when the input remains the same. To make each hash password unique, random data is added to the input of the hash function [97].

In an existing digitalized system, portability of patient information from one health-care facility to other is very less which affects providing quality of care to the patients. The Aadhar number allows the identification of an individual, which is unique to that individual. Attaching the Aadhar number to a patient's health information gives way for easy access and portability, which in turn allows for a more patient-specific approach to treatment. Aadhar authentication system provides multi-factor authentication, wherein Aadhar number with other attributes such as biometrics, iris scans , demographic data and OTP are submitted on-line to the Aadhar authentication server for user verification on the basis of provided information. In addition to strengthen user authentication, OTP or PIN is also be used in conjunction with Aadhaar authentication.

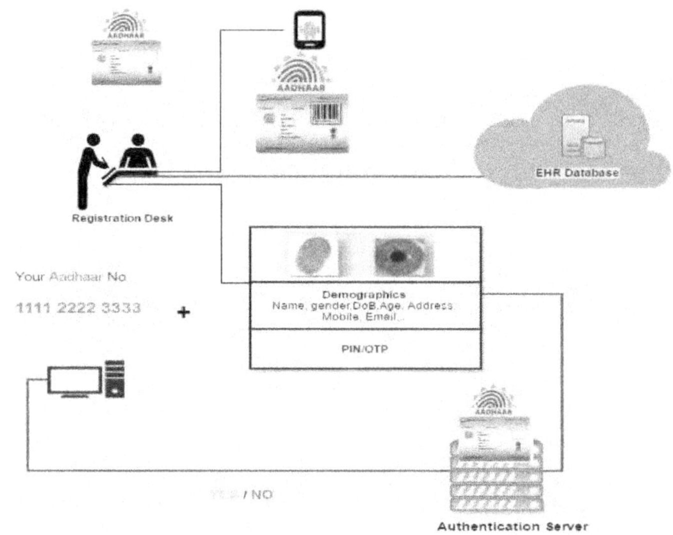

Figure 4.6: Aadhar enabled EHR system

The proposed model for Aadhar enabled EHR is as shown in Figure 4.6. During patient registration or during hospital visit, ID proof like Aadhar Number is used as a universal patient identifier. Healthcare user authentication is archived at database level through valid user name and password. If the end user is not logged in and requests a page that requires authenticated access, he or she is redirected to a login page which will prompt them to supply the respective credentials typically a user name and password. The cloud server checks the credentials against a user store of the database for validation as shown in Algorithm 2. When the user is registering for the first time, the system will send a unique code to the respective user's mobile. After the pass code is verified, the user is granted access to the system. A 2048-bit PKI encryption is used by the model which, when combined with HMAC, ensures that the data in the cloud is not altered in any way.

Algorithm 2 User Authentication

```
procedure AUTHENTICATION(User ID, Password)
    A=Fetch stored Hash value from the Cloud Password file
    B=Compute Hash value of the entered password
    if ( thenA==B)
        access grant
    else
        Deny the access
```

4.3.2 Authorization model for cloud based EHR

Authorization services plays a vital role in enhancing access role of cloud resources. It is an identity service which comprises of policy management, role management, and role-based access control mechanism to manage access to the applications [98]. Cloud based EHR system with OAuth service for authorization is shown in Figure 4.7. Authorization model contains four rows for representing user, user application (browser), Authorization Server(AS) and resource server. Healthcare users own the resources which is stored in the remote server or remote database. If end user wants to access the resources, first he/she has to enter the credentials through web browser interface. The request is sent to the Authorization Server (AS) for a particular record or resource. The request is composed of data such as an 'Options' field , ID_c (client ID), Timestamps for request validation, nonce and service. The AS verifies the user's credentials following which it returns an encrypted token that permits the user to make subsequent requests to the cloud server and gain read-write access to the authorized resources. The results which are returned to the user are encrypted with a key derived from the user's password K_{cv}.

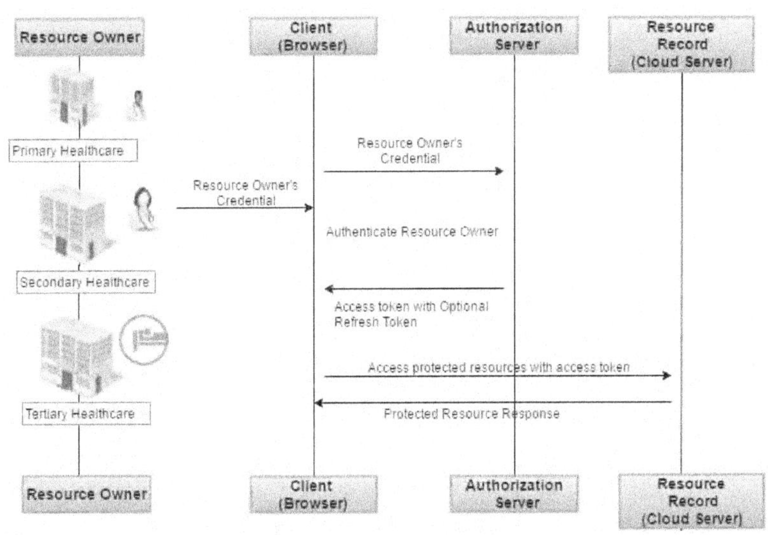

Figure 4.7: Authorization Model for healthcare system

To summarize, the client receives an authorization token from the AS once their credentials have been validated and verified. In order to gain access to any resource from the cloud server, the client will embed the authorization token in every request to fetch, create or modify a record in the server. The requested resource once retrieved is then shown on the client's web interface in the browser. Thus, without the token, the user's request to access a record will be rejected. Table 4.5 enlists the data components required for the authorization process and the cryptographic syntax is illustrated in equation 4.2.

Table 4.5: Elements of authentication service

Options	Designation	Healthcare level	Rural	Urban	Year of experience
Idc	Aadhar number				
Times	Used by the client to request the following time setting in the ticket. From: Start time for the requested ticket Till:Expiration time for the requested ticket rtime: requested renew time				
Nonce	A random value assure that the response is fresh				

$$
\begin{aligned}
C \rightarrow AS &: Options \parallel Id_c \parallel Times \parallel Nonce \parallel Service \\
AS \rightarrow C &: Options \parallel E[Id_c \parallel Times \parallel Nonce \parallel K_{c,v}] \\
AS \rightarrow V &: Service \parallel Options \parallel Times \parallel K_{c,v} \\
C \rightarrow V &: E[K_{c,v} \parallel Message] \\
V \rightarrow C &: Nonce
\end{aligned}
\tag{4.2}
$$

4.3.2.1 Procedure to access EHR record using using OAuth 2.0

Using OAuth method user can access the EHR data over the network on the basis of authorization code and token. Figure 4.8 shows the step by step procedure to access the health record through authorization process. If health-care professionals wants to access health data of patient, user has to click the authorize button. Once he/she clicks, the request will go to EHR server where OAuth server is running. Once server receives the request, web application shows list of patient. Healthcare professional are given provision to select the patient from the list for accessing the record. For performing this, the user has to click the button named "Yes, I Authorize The Request" as shown in the Figure 4.8 in step 2.

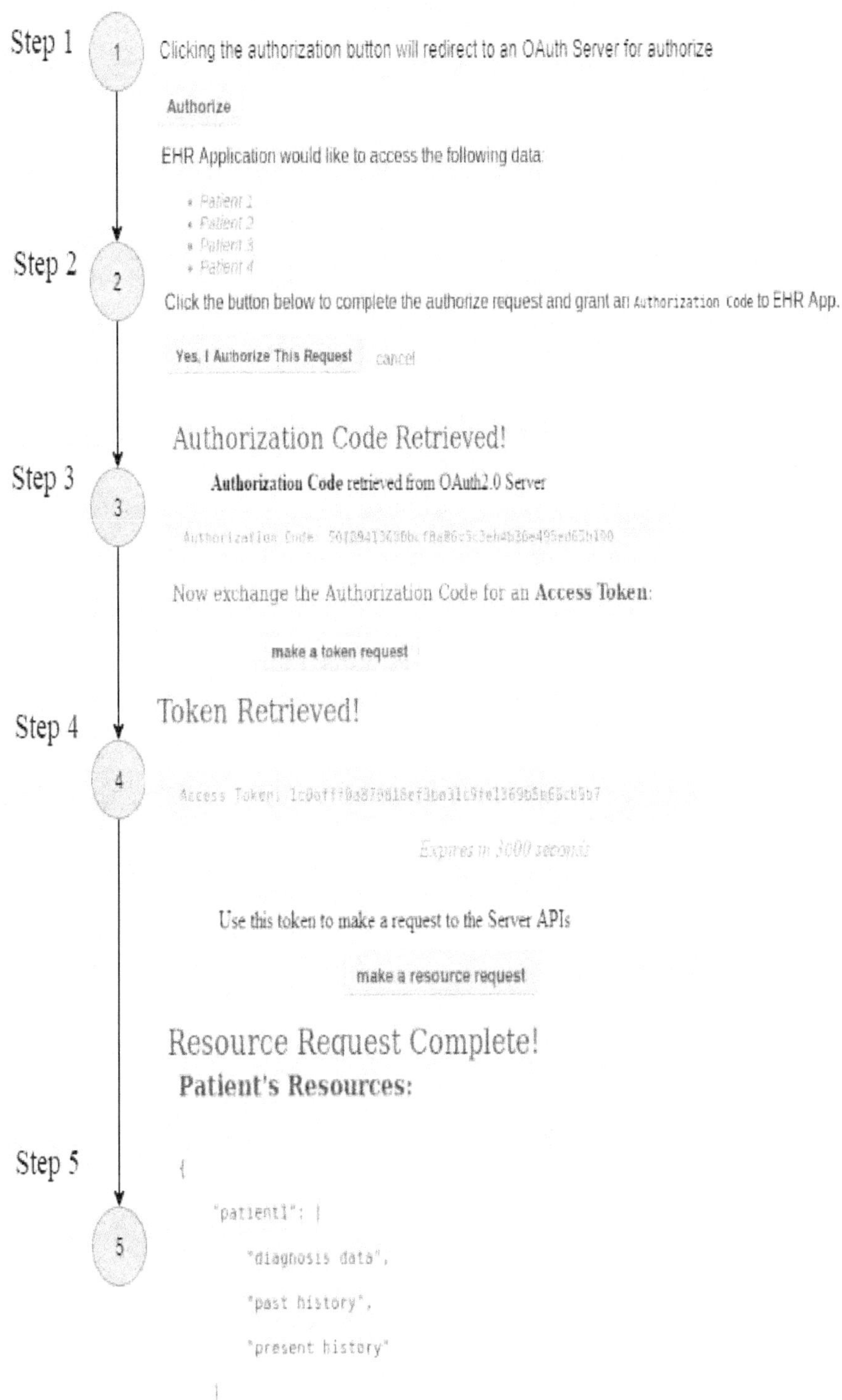

Figure 4.8: Step by step procedure to access the health record

Authorization code is generated by the server and code is exchanged with client For example the authorization Code generated is 675931b9e0137277c2891a64682d025a

96850a95. With this code, the user can access the token by sending authorization code to server. As soon as server receives the request, access token is generated with expiration time of 3600 seconds. For example, suppose access token received is d3291f6a495542d6029ba2f225a2374 d4cbde74d. Using this token user makes request for the EHR data and user is redirected to the EHR data. User can access the health data of patient till token expiries. Example of EHR data accessed by user is shown in Figure 4.8 in step 5.

4.3.3 Role Based Access Control (RBAC)

The RBAC is one of the best method for highly complex and huge management system. In this method, access is confined to certain users based on their roles and special privileges given to roles in some cases. Table 4.6 and Table 4.7 shows the functionalities of the web based EHR system and healthcare user's role and permissions.

Table 4.6: Roles of healthcare users

Healthcare Users	Permissions	Security Module	EHR Standards
Receptionist	1. Login 2. View EHR Record: Demographic Data Appointment 3. Add/Update EHR Record Demographic Data Only Appointment	Authentication Authorization Automatic Logoff Audit Log SHA-256 Oauth2.0	Minimum Data Set (MDS)
Patient	1. Login 2. View EHR Record: Demographic Data Prescription Data	Authentication Authorization Automatic Logoff Audit Log SHA-256 Oauth2.0	MDS
Nurse	1. Login 2. View EHR Record: Patient demographic data Patient history 3. Add/Update EHR Record Basic Vital signs of the patient	Authentication Authorization Automatic Logoff Audit Log SHA-256 Oauth2.0 RSA	MDS

The permissions are defined in the permission column where the healthcare users can access the functionalities of the system. Each role has got its own significance, and privacy of each user is preserved. There is another role called administrator who has full access of the system.

Table 4.7: Roles of healthcare users

Healthcare Users	Permissions	Security Module	EHR Standards
Doctor	1. Login 2. View EHR Record Demographic Data Patient past history Present History Basic Vital Sign Allergy status 3. Add/Update EHR Record Diagnosis Details Laboratory Test Radiology Test Prescription	Authentication Authorization Automatic Logoff Audit Log SHA-256 Oauth2.0 RSA Access Control	ICD-10 NDC
Lab Technician	1.Login 2. View EHR record 3. View Test Report 4.Update EHR Record Add/Update Test Report	Authentication Authorization Automatic Logoff Audit Log SHA-256 Oauth2.0 RSA	LOINC
Radiologist	1.Login 2. View EHR record 3. View Radiology Report 4.Update EHR Record Add/Update Radiology Report	Authentication Authorization Automatic Logoff Audit Log SHA-256 Oauth2.0 RSA	DICOM
Pharmacist	1.Login 2. View Prescription details 3. Select the prescription 4.Update EHR Record Add/Update selected medication Report	Authentication Authorization Automatic Logoff Audit Log SHA-256 Oauth2.0 RSA	NDC

4.3.4 Attribute Based Encryption(ABE) for secure storage of EHR system

Cloud based EHR allows centralized storage of patient health information which generally improves the process of storage and retrieval of the patient data in health information system. Due to the storage of sensitive data in 3rd-party storage, there is a need to encrypt data stored at the cloud server sites. To attain secure and scalable access control to EHR data, ABE[99] techniques are used. In this technique different set of attributes are used to encrypt health data which mainly reduces key distribution complexity in an cloud platform. ABE enables guaranteed patient privacy and allows dynamic modification of access policies when patient record is transferred from one healthcare facility to other. In the proposed system, the patient is the owner of the record, where he/she can control access by giving authority to access their health records. When healthcare service provider has to access the record he/she needs to provide a proper attribute to satisfy the access policy condition to decrypt the EHR record. Those who are unable to provide necessary details are prevented from decrypting the records.

In the context of security and information access, revocation is a process where a user who was previously granted access to a particular data loses his privilege of accessing it in the future. This may be due to reasons such as the data item getting modified or new data getting generated in its place. The primary purpose of revocation is to ensure that data owners continue to maintain control over certain information despite giving the required access to the recipients. On demand revocation policy enhances the security by blocking the users from using the same attributes to access the future EHR records. Each time attribute revocation is necessary for the record access, if the attribute mismatch results in user request to access a record will be rejected.

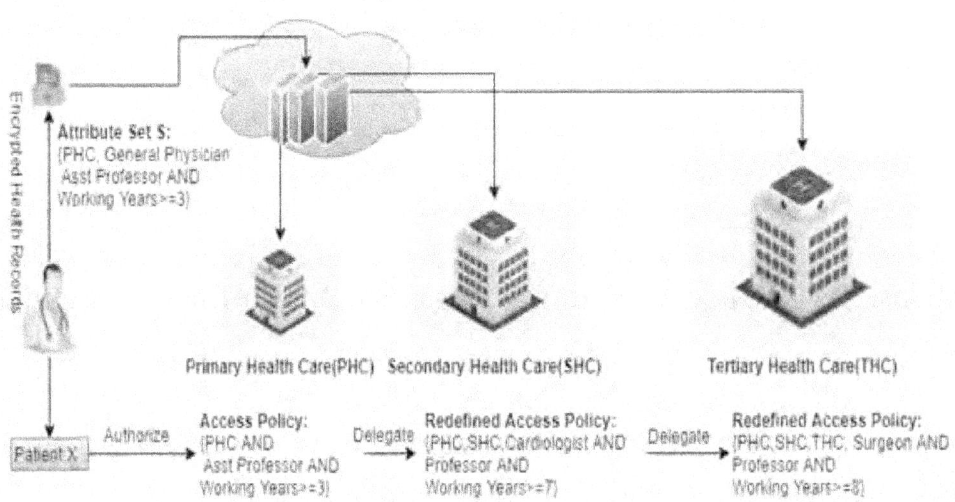

Figure 4.9: Attribute based encryption for EHR system

The proposed framework for patient centric, secure and scalable EHR sharing on cloud storage is depicted in Figure 4.9.

The attributes such as Healthcare Institute Name, Health Department, Doctor

Designation, Work Experience are considered as the common attributes in each level of healthcare system. Assume that patient health record from the PHC is encrypted with the following attributes: Primary healthcare, General medicine AND Assistant Professor AND Working Years ≥ 3, which is denoted by A. If the same patient needs consultation by a Senior doctor of cardiology department of SHC, then instead of sharing secret key with all health care professionals, current doctor from PHC can give access policies having following attributes: Primary health Care, Secondary Health Care AND Cardiologist AND Professor AND Working Years ≥ 7 denoted by $A^{|}$ and further generates a secret key S_{KA} using it. With this key pair A and S_{KA}, higher level health-care professional (SHC)can access patient data. This process of allowing access to other health-care center doctors is referred as redefining access policy. Similarly, when the same patient needs to be referred to a surgeon working in a higher level hospital such as THC, current doctor in SHC can give access policies having following attributes Primary health Care, Secondary Health Care AND Surgeon AND Professor AND Working Years ≥ 8 denoted by $A^{||}$ and further generates a secret key S_{KA} using it. Attribute based encryption ensures that security and privacy of sensitive patient data is protected from unauthorized users.

4.3.5 Attribute-Based Access Control

Patient health information sharing and exchange across different healthcare levels require managing federated identities and various access control policies. Attribute-Based Access Control (ABAC), also known as Policy Based Access Control (PBAC) refers to defining access policies and authorizing access to resources and data based on certain attributes defined by the organizations [100]. In ABAC multiple attributes are considered to grant access to the resource. Attributes associated with the user are context of the request, subject of the information, attendance timestamps, role hierarchy, work experience, and consent by the patient. Attributes can vary from organization to organization. In contrast to Role Based Access Control (RBAC) which provides access based on predefined roles, ABAC allows organizations to set a complex set of Boolean logic and conditions to ensure that access is provided if and only if all the conditions have been satisfied.

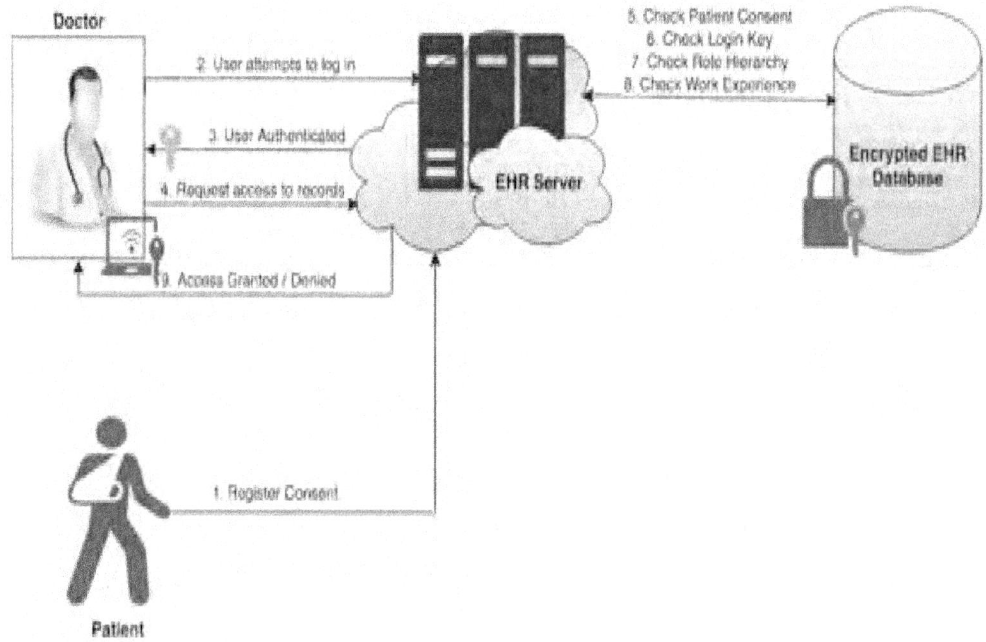

Figure 4.10: Sample work-flow of access control

Figure 4.10 illustrates a sample work flow demonstrating ABAC in healthcare organizations. For example if healthcare users from hospital A wants to access the patients health record, the access is granted only if the patient or his family has registered and given their consent. Now hospital has been authorized to give access to trusted users. Once a hospital has been authorized, the hospital policies come into play. If a user satisfies all the conditions and policies set by the healthcare organization that they are working in, then they are granted access to that patients records.

ABAC given in Algorithm 3 is mainly used for granting access to the healthcare users. In the first step the patient consent is received by the hospital authority and the value YES is used to continue the loop. Web application requests healthcare users to enter the login credentials and department details. In this example healthcare user is belonging to CARDIOLOGY department. Next control moves to the role hierarchy attribute. It is a numeric value that refers to their designation in the organization as shown in Table 4.8. Work experience is also a numeric value that refers to the number of months the user has worked in the organization.

Algorithm 3 Accessing and updating health information of the patient

```
if ((user.getHospital().getPatientConsent())=='YES') then
    if ((user.getLoginKey().isValid()) then
        if (user.getDepartment()=='CARDIOLOGY') then
            if (user.getRoleHierarchy()≥min (task.getAuthorizedRoleHierarchies())
AND user.getWorkExperience() ≥ task.getRequiredWorkExperience()) then
                Access Granted -Update Patient Records
            else
                Access denied - Contact Supervisor
        else
            Access denied - Contact the concerned department
    else
        Access denied - Validity of login key has expired
else
    Access denied - Contact concerned patient for the hospital authorization
```

Table 4.8: Access role of healthcare users

Role Hierarchy	1	2	3	4	5	6	7
Role	Chief of Medical Staff	Department Chief	Sr Resident Doctor	Jr Resident Doctor	On Call Doctor	Medical Interns	Medical Students

Patient record transfer with read and write access control is depicted in Figure 4.11. During record transfer, patient needs to approve the transfer request. After confirmation from the patient, referral hospital admin approves a transfer request sent by the referred hospital admin. The healthcare professional at referral level will have access to the complete patient health information. This transfers the read write access control to the SHC to access the patient record. Now the PHC will have only read access. In order to know, with whom the read write access control lies, an extra attribute hospital ID(h_id) is used. This is a list which indicates the path of transfer of the patient record. Whenever patient record is transfered to other hospital, the h_id is appended to the list. Thus the last h_id is the list indicates the hospital that possesses the read-write access control.

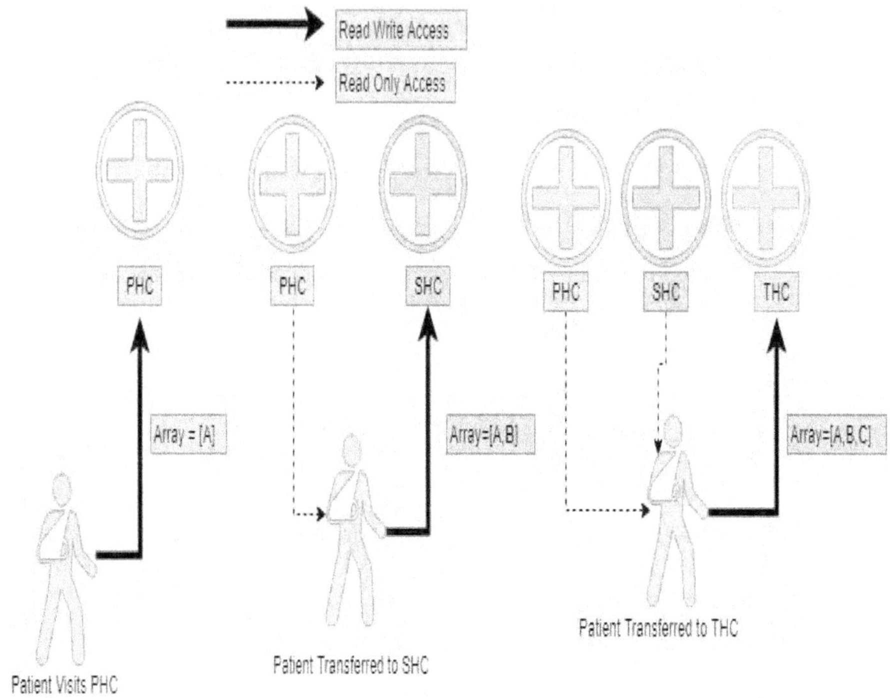

Figure 4.11: Record transfer with read/write access control

The patient record transfer request process is discussed in the sequence diagram as shown in the Figure 4.12. When patient moves from one hospital care to other level, to access patient record in other health care level, patient confirmation is needed. To enable the transfer, host hospital Admin needs to send a confirmation request to authorized patient who need a transfer. After confirmation from the patient the health data of the patient is available for guest hospital health care professionals. The Role-Based Access Control (RBAC) of the system permit the referral level doctor to view the patient data without any difficulty with valid authentication. This record transfer approach allows the doctors at the referral level to get access to the complete patient demographic data, patient history, diagnosis information (CT scanners, MRI scanners) and pharmaceutical information which is stored in the EHR database. Here, the referral doctor can view the patient data and if required, update the diagnostic and therapeutic details of the patient. If the patient denies the transfer request the transfer request mode is automatically switched to disable mode and patient data is accessible only at the referred hospital.

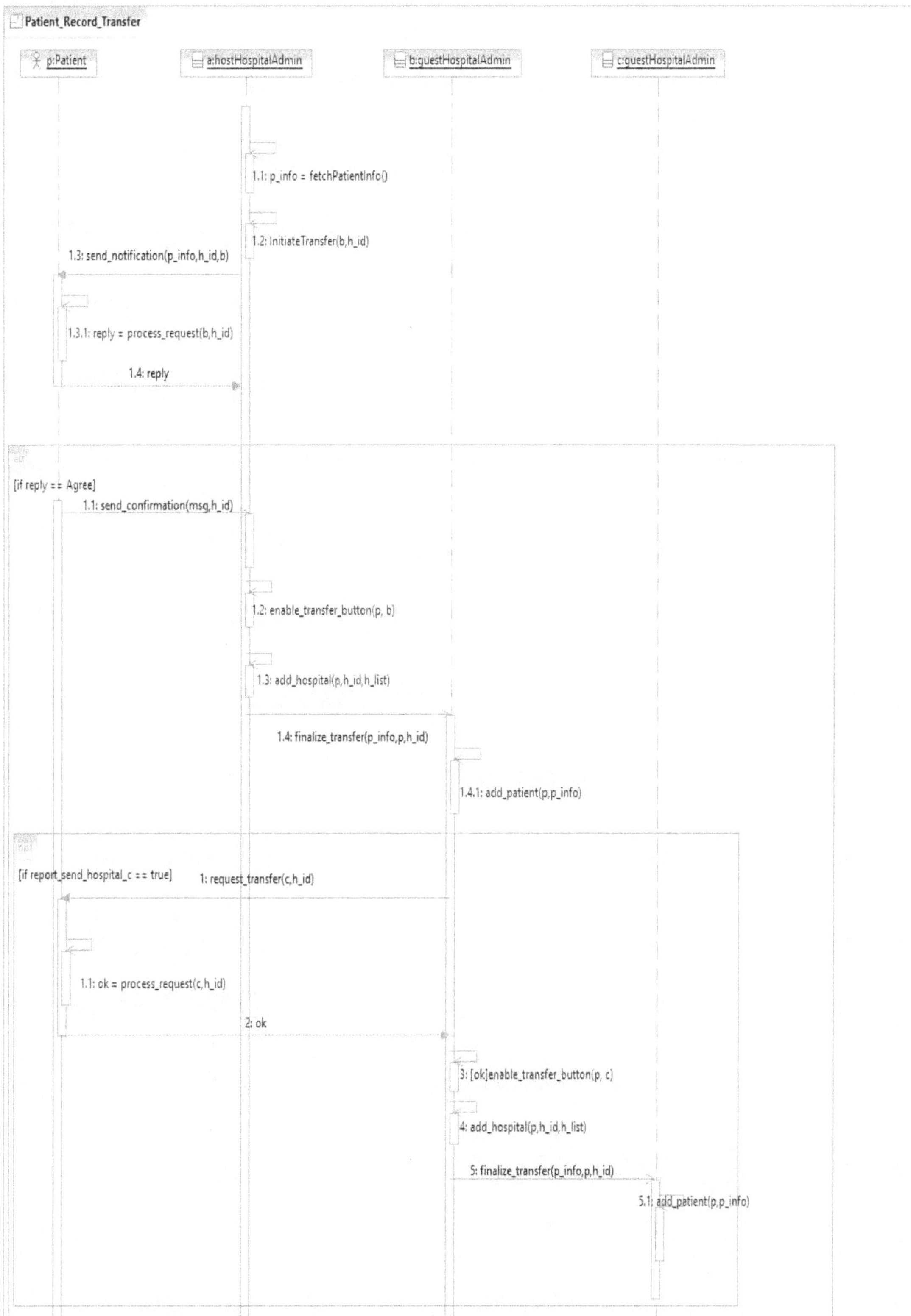

Figure 4.12: Patient record transfer from one healthcare level to other

4.3.6 Digital Signature

A digital signature is the one of the main aspect of ensuring security and integrity of the patient data[101]. In healthcare system, medical forms and prescription details of the patient needs to be signed digitally to prevent from the unauthorized access, tampering and forging a signature. Doctor from PHC can sign a document using digital signature procedure as shown in Figure 4.13.

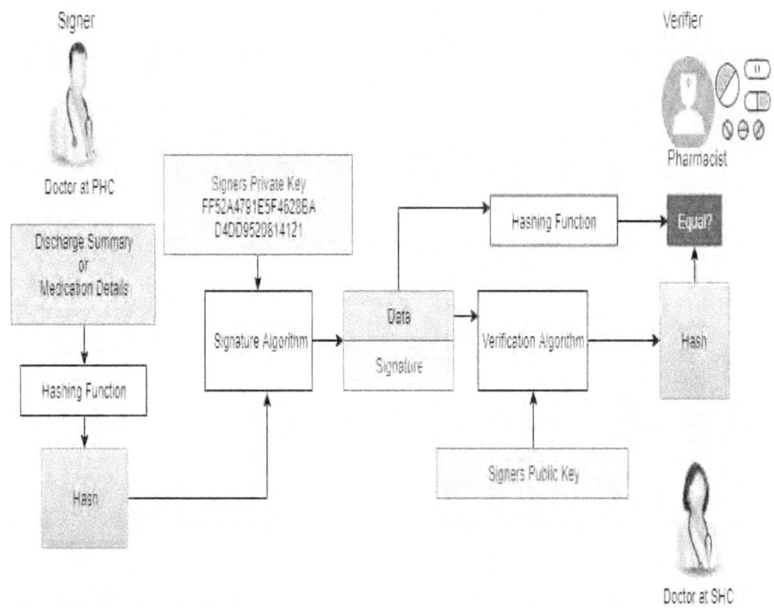

Figure 4.13: Digitally Signed Prescription Process

The input to the message is doctors private key and medical form. Any other healthcare user from the other level can confirm the signature with the help of verification algorithm, which requires received medical form and PHC doctor's public key. The implementation details of digital signature is shown in Algorithm 4. Doctors were able to authenticate all their issued documents with personal signature online within a few seconds using Docusign REST-API. This enhanced the functioning of EHR in this model as not being able to approve documents with signature was a major drawback of previous traditional system.

Algorithm 4 Authenticate all issued documents

```
procedure ( )
    Get data from HTML form
    send_dict=API references+all data in json format
    set_headers=API_headers
    r=requests.post(API_url,headers=headers,json=send_dict)
    res=json.loads(r.text)
    db.insert(Doctor ID: doc_id, Envelop ID: res[envelopeId],Patient Aadhar: aadhar)
```

4.3.7 Dual Database for Securing the Patient Data

The security of EHR system is the main concern in all levels of healthcare system. All patient record is in the cloud and can be accessed from any where at any time. In the public cloud environment, once data enters into the cloud service provider premises complete control will transfer to them. The security model for storing and retrieving sensitive personal information of the patient are depicted in Figure. 4.14.

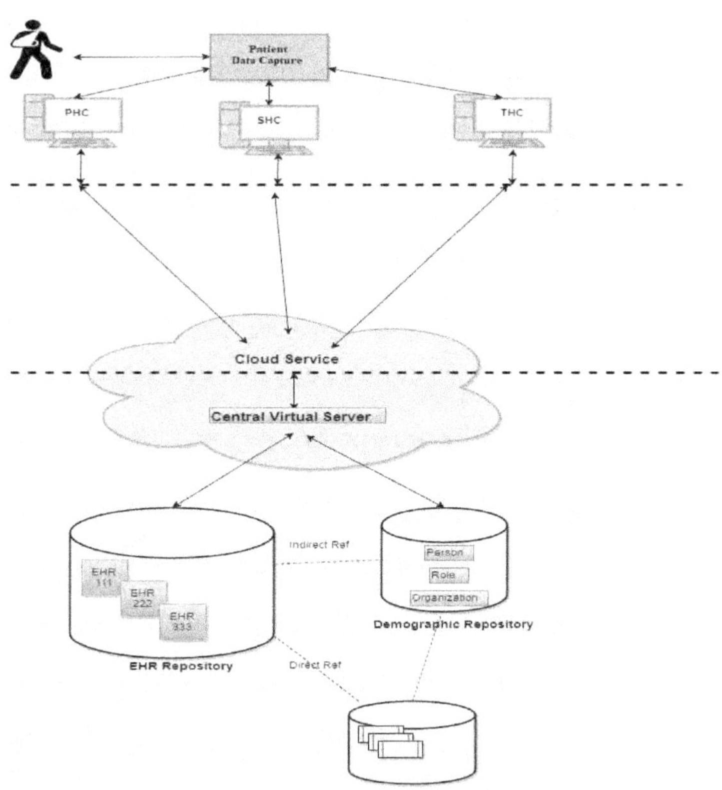

Figure 4.14: Security model for storing sensitive information

The model consists of central health record server in communication with a medical record repository or database. The medical record of the patient is sensitive, which includes biometric information, lab examination result, psychological and mental health condition, family history, allergies, medications taken, medical conditions, past medical treatments, and diseases.

The Health Information Management (HIM) system needs to ensure that data is available when needed and needs to safeguard the health information system from unauthorized user. The medical record maintained in the repository or database is identified using patient identification number.

In this model of architecture, health record of the patient called EHR is categorized into demographic data and health repository. Two different database is used to store these data separately. One database handles all health related information and another for demographic details(biometric information etc.). Health record database need not include personal identification of the patient (patient name). Instead patient identification or biometric information is stored in the demographic database. Patient health information stored in the health records may be identified using one or more identifiers associated with the patient. Both the database are associated or linked by using an alphanumeric pass-code. Using this pass-code medical records can cross-reference with demographic data(biometric). In hospital, large amount of data is collected every day and recorded in the EHR database. The EHR data can be used in two different ways by the users of the system. Healthcare professionals use EHR for health related functionalities and scientist and researcher mainly to discover the new knowledge.

4.3.7.1 Managing keys and encryption

The dual database model ensure security by using two different cloud databases. Encrypted health information of the patient is stored in one database and key related information to perform encryption and decryption are stored in the other database. With this approach path for obtaining sensitive information to the attacker is blocked. The data will be stored on the cloud through cloud gateway. The interface between gateway and cloud database is shown in 4.15

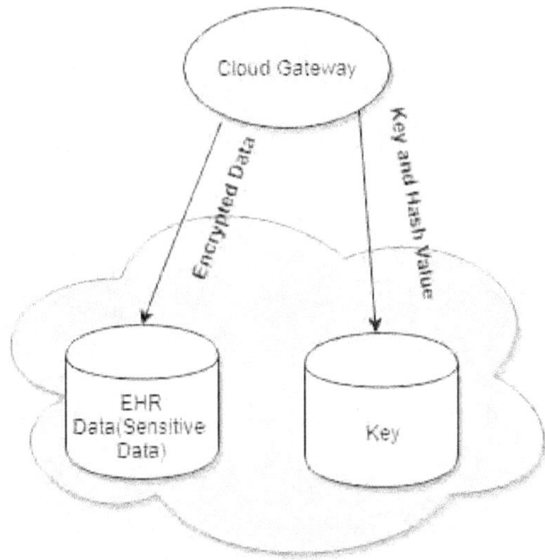

Figure 4.15: Interface between gateway and VM Database

The data is encrypted at the gateway using key from key database and stored in EHR database with timestamps. Before storing data into the database, encryption is applied at the gateway to maintain confidentiality of the patient data. Pubic key cryptosystem is a technology used to encrypt the original message into encrypted format. The method used for encryption is the RSA as shown in Algorithm 5. RSA uses two different keys, one key is used as public key other used as private key.

Algorithm 5 Key generation ,encryption and decryption using RSA

procedure ENCRYPTION(EHR Data)
    m=EHR data to be encrypted
    Generate two large prime numbers p and q
    Calculate Modulus n=p*q
    Calculate quotient value of the Eulers function for n is Φ = (p - 1)(q - 1)
    Find the random number public key e such that 1 < e < Φ and gcd(e, Φ) should equal 1
    Calculate private key d such that d = e^{1} mod Φ(n)
    Assign public key as (n,e)
    Assign private key as (n,d)
    Compute the Cipher text C = m^{e} mod(n).
procedure DECRYPTION(Encrypted EHR Data)
    Decrypt to get plain text M = $c^{d} mod(n)$.

The encrypted data is hashed to perform integrity, which prevent the intruders from modification. Equation 4.3 is used to compute the hash value.

$$H_i\{m\} = Hash(E_i(m)) \parallel Key_i \tag{4.3}$$

where $H_i(m)$ is the hash value of the ith data, E_i is the encrypted data and Key_i is the key used for encrypting the ith data. So $H_i(m)$ supports in maintaining the integrity of the encrypted message and also the key used to execute encryption operation. During retrieval, the hash is computed to check the integrity of the stored data. If satisfied, the corresponding key is used to perform the decryption operation. This decrypted information is passed on to the service cloud. Later service cloud displays the decrypted information to the end user in the required format. The dual cloud database model protects the sensitive information of the patient by maintaining the confidentiality and integrity of the data. So, healthcare users can access required information from the health information system with secure manner.

4.4 Secure Deployment Architecture for EHR system

Figure 4.16 shows the proposed secure deployment architecture for EHR System. The first tier is user interface also called presentation tier. With this level, healthcare professional gets access to the healthcare system. In order to provide security, authentication model SSO with salting methods is adopted. The SSO and salting methods in user interface improves the security, and it makes it difficult to access the resources for unauthorized users or attackers. The second one is the functional process or business logic which can be secured using secure access protocol such as HTTPS and SSL. The last tier is the data storage tier and the information stored in the cloud data base is protected using OAuth, RBAC, ABE, ABAC techniques. Authorization module OAuth ensures that the patient's data can be accessed securely without compromising the patient's or the user's account credentials. To protect the data from the unauthorized access, encryption techniques RSA and ABE are being used to store data at cloud premises. RBAC and ABAC ensures in defining access policies and authorizing access to the resources.

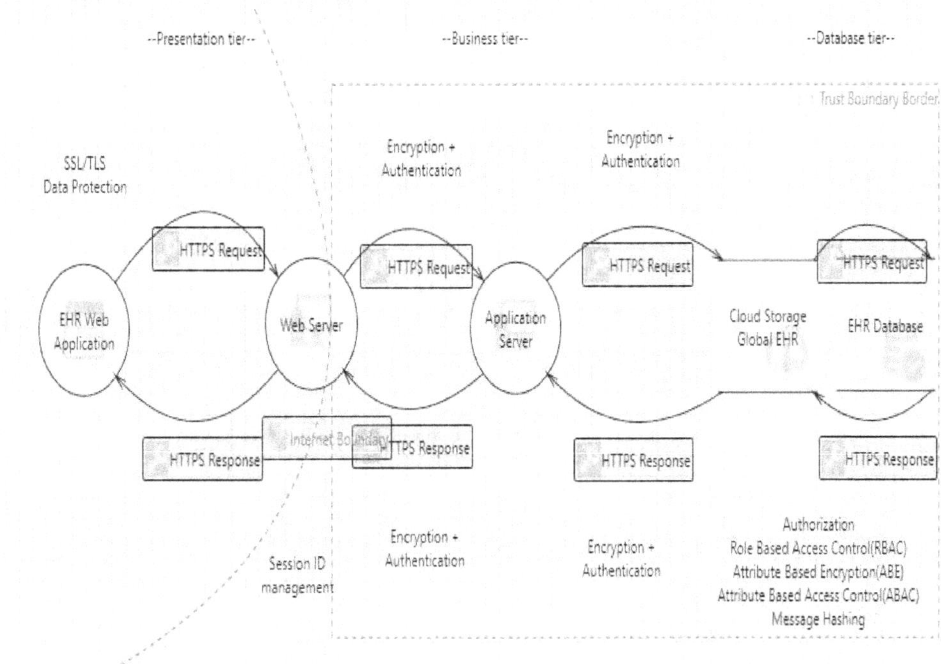

Figure 4.16: Secure architecture for EHR system

The security framework can be easily adopted across different levels of healthcare system and also ensures Confidentiality, Integrity, and Availability (CIA) of the data.

Summary

The Security architecture for the developed cloud based EHR and various attacks and vulnerabilities that has been identified using Microsoft threat modeling tool is discussed in this chapter.Security algorithms to mitigate the threats and vulnerabilities in the system are explained. The subsequent chapter discusses the deployment of the developed EHR system in the cloud based platform.

Chapter 5

Deployment of Developed EHR system in Cloud

5.1 Introduction

Cloud computing provides the flexibility to the users to access files, applications and services over the network. The term cloud computing is new technology for most of the users until late 2007. The term ‘cloud first achieved significant attention when Amazon released Elastic Cloud Compute also popular as EC2 [102]. Cloud computing overcomes the issues that are faced by the classical data centers by using virtualization. Virtualization abstracts physical resources which ends up appearing as logical resources. It can be implemented at different layers such as compute, storage, network and even the application layers. Cloud service providers offer their services categorized as several fundamental service models [103]. But three among them are considered fundamental:

- **Software as a Service(SaaS)**: SaaS offers a complete application as a service by running an instance of it on the cloud.
- **Platform as a Service(PaaS)**: PaaS comprises of an environment wherein cloud applications are developed and provisioned. The environment itself is made available to the end user.
- **Infrastructure as a Service(IaaS)**: IaaS facilitates the delivery of standardized services in the form of basic storage and computing capabilities to

ends user over the network.

Apart from major service models, there are many other models including network, storage, security, data, database, information, API, back end, process, integration and management as a service model to perform different functionalities in the cloud computing environment. There are many ways in which the cloud can be deployed. The main factor for choosing the model of deployment is depend on requirement of the client or organization. In general, there are four fundamental models of deployment of cloud:

- **Private Cloud**: Virtualized resources are shared privately within the organization and it has provision to access and manage the computing resource within the network boundary of organization firewall.
- **Public Cloud**: A public cloud delivers remote cloud services over the internet.
- **Hybrid Cloud**: A hybrid cloud combines two or more distinct cloud environments (public or private) from different providers.
- **Community Cloud**: A community cloud provides an infrastructure to satisfy a set of common requirements of the related stakeholders using semi-private clouds.

Cloud Computing has many advantages over the traditional dedicated systems. Virtualization is responsible for the increasing demand of cloud computing. The advantages provided by cloud computing include tremendous flexibility, significant cost reductions due to low infrastructure investments and the pay-per-use facility of cloud, scalability as cloud computing provides an option to scale the resources whenever required and elasticity as allocation and de-allocation of resources is automatic in an cloud environment. Figure 5.1 depicts the organization of chapters to explain deployment infrastructure for developed EHR system. The first section contains the open source infrastructure to manage the patient data and second section present the public cloud infrastructure setup.

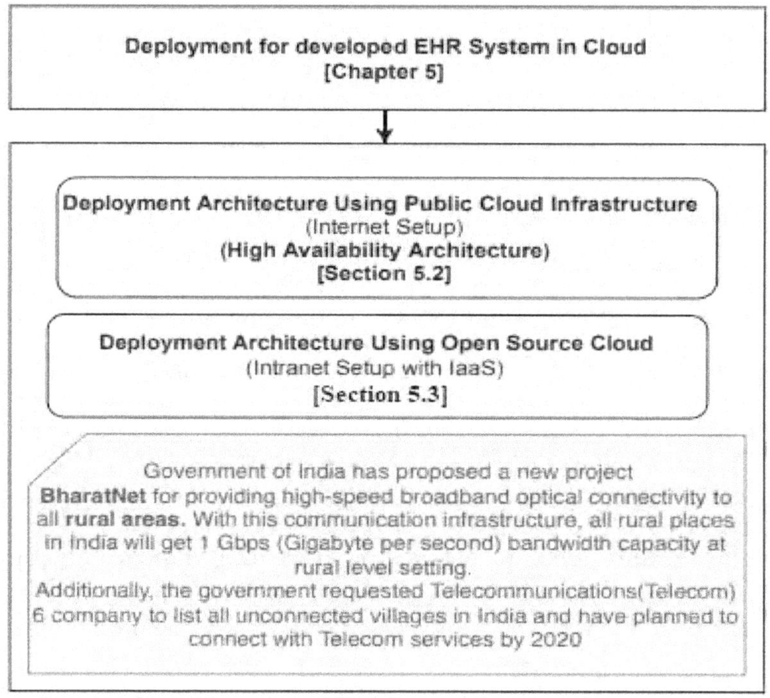

Figure 5.1: Chapter Organization

5.2 Deployment architecture using Public cloud infrastructure

Public cloud infrastructure enables to manage patient health information from primary, secondary and tertiary of different healthcare levels. Here, the primary level is again divided into sub-center, primary health center, and community health center. All the data captured from each healthcare setting is finally stored in global cloud database called EHR database. The Amazon cloud service infrastructure is used to build high availability cloud infrastructure for PHC, SHC and THC.

5.2.1 High availability architecture for EHR model

To build a high availability architecture the services of Amazon Lightsail are used [104]. Cloud set up is created using virtual private servers. High availability architecture of EHR system is shown in Figure 5.2. It is substantial and expected to operate continuously without failure for a long amount of time. Mainly, fault tolerance characteristics features of the architecture allows to remain in operation even if some of the component used to build the system fail. Major building blocks of high availability architecture is as follows:

- Healthcare user
- Load Balancer
- Availability Zone
- Snapshot-Replace instance and Re-attach volume
- Replication Layer-Moving data from one to another

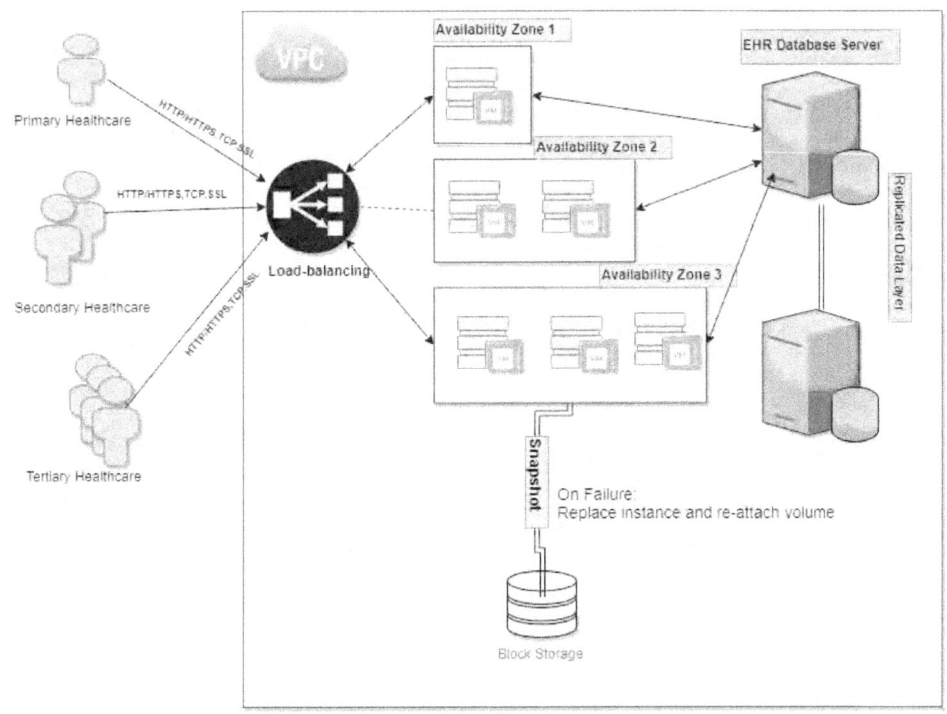

Figure 5.2: High Availability architecture for EHR system

The health care users request health care services from the EHR system and receive response from the EHR system. With load-balancer features, availability of the system increases by distributing the load among the zones [105]. The request always move to the healthy running instances instead of going to the unavailability zone. Load-balancing facility allows automatic distribution of incoming application traffic among multiple instances of the cloud storage. Also load balancer is configured to handle encrypted (https) traffic, session persistence, health checking, and more. The availability zone is another major component of the architecture where it is hosted in multiple locations as a zone. Snapshot feature provides the backup of the instances during the instance failure. The backup layer protects from failure. The EHR data is stored in global database. Health care users are allowed to perform read, write and update the health data which is stored in the primary EHR database. The data replication layer provides the functionality of switching data between primary and secondary database. In case if primary database fails still user can access data from the secondary database. When healthcare users grows, user can add block storage disks and attach them to the created instance for adding more users to the system.

Providing security of data is categorized into data at rest and data in transit. So, SSL/TLS assist sensitive data in transit by encrypting the data and authentication using SSL/TLS certificates to establish secure connections between browsers and EHR applications. Created instance or virtual machine has images with base operating systems, development stacks like LAMP, and SQL Server Express. To access the instance of the virtual machine, private key is created by configuring PuTTYgen, after using the key Open PuTTY by entering the public IP address instance in the putty management page. It enables virtual machine with proper IP address.

The Bitnami LAMP stack services in amazon simplifies the development and deployment of EHR applications [106]. It includes ready-to-run versions of Apache, MySQL, PHP, and phpMyAdmin, as well as the other software required to run each of those components. PhpMyAdmin services are enabled in the virtual machine instance.

Cloud-enabled standard EHR application has multiple functionalities, the high availability cloud architecture bringing it all under the one roof under the AWS cloud ecosystem. Patient health data recording and reporting are done using standard codings such as SNOMED-CT, LOINC, DICOM and NDC. The healthcare standard and medical coding with cloud database enable connectivity between primary, secondary and tertiary levels to make patient health data available all-time when required for his/her care.

5.3 Deployment architecture using Open Source cloud framework

Advancement in IT technology in healthcare sector has made it possible to maintain and manage the patient data in digital form in all levels of healthcare system. Cloud computing has emerged as a better solution for healthcare IT sector to manage patient data. Therefore, rural healthcare organizations should move towards building their own private cloud infrastructure which could be an excellent solution for the country's needs to have improved healthcare in rural areas.

As far as infrastructure provisioning is considered for cloud, there are only two major players OpenStack and Eucalyptus [107]. Infrastructure provisioning involves a provisioning tool to supply virtualized resources on-demand. Being API-compatible with Amazons EC2 cloud service, Eucalyptus can be used by any company to build a compatible test lab for the verification of their requirements. After being registered as an Amazon customer, companies can then use Eucalyptus to perform the required development and testing activities before pushing to the cloud of their choice. OpenStack offers self-service clouds in a secure, private test lab before moving to either private or public cloud. It also provides insights on security issues related to cloud computing and IaaS. OpenStack has been recognized as one of the most significant advances in cloud computing till date. OpenStack offers much more flexibility than Eucalyptus. OpenStack lacks in one thing compared to Eucalyptus, the tie-up with the leading cloud-provider, Amazon.

So, both the platforms have their benefits and drawbacks. OpenStack does not offer compatibility with EC2 APIs but offers much more flexibility and customization to cloud projects. Eucalyptus does not have much customization and flexibility but offers compatibility with AWS. Which is the first open source developed in early 2008 [108] and it uses AWS API-compatible platform for developing varieties of cloud models. Further Eucalyptus has an open-source version and commercial version whereas OpenStack is completely open-source. It completely depends on the requirement of the prospective user. If the user considers AWS for cloud services, then Eucalyptus is the best choice.

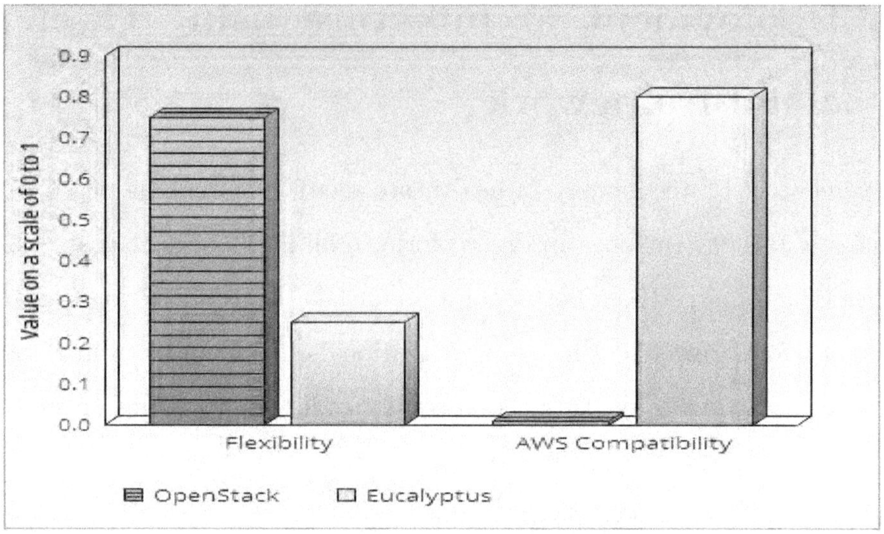

Figure 5.3: OpenStack vs Eucalyptus

Open Source Cloud frameworks are adopted and it has been found that Eucalyptus is ideal for Electronic Health Records. Using Eucalyptus user can build their own private and hybrid cloud computing environment. Figure 5.4 shows the architecture model which describes one simple application of IaaS on private cloud using EUCALYPTUS. For this purpose, Faststart model of installation was used with Eucalyptus version 3.4.2. All the necessary setup was done and an instance was launched with the custom Ubuntu Karmic image. Then by accessing the instance from terminal using ssh, proper network settings like proxy settings, DNS servers were configured and the network connectivity provisioned to the instance. Using this connection, tomcat6 was installed on the instances of Ubuntu operating system.

Eucalyptus is a private cloud framework where patient sensitive record can be stored. Eucalyptus instances or virtual machines can be deployed and accessed privately. However with public IP address, user can access the VM instances through Internet. Although limiting the access of private cloud to intranet is vital as it provides security, it may be necessary to provide access to the private cloud from outside the intranet (Internet). Providing access means providing access to one or some of the VMs that are running on the private cloud. These VMs can be accessed using ssh which uses port 22.

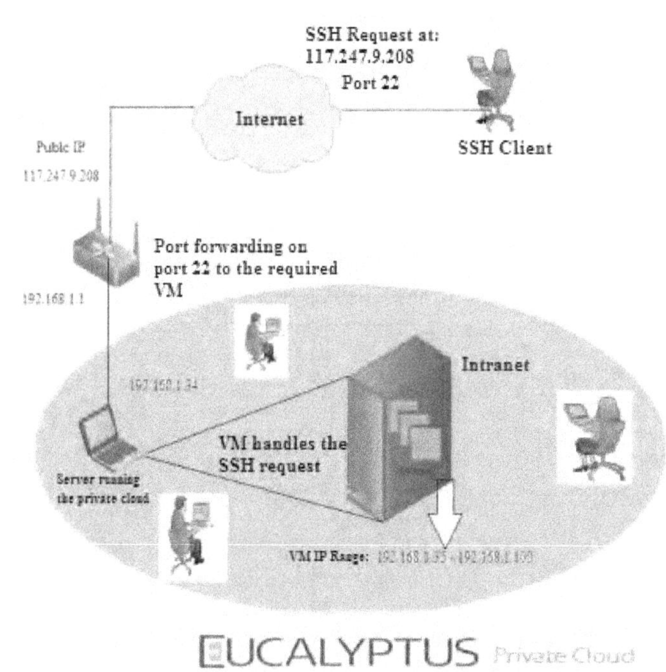

Figure 5.4: Deployment architecture using eucalyptus

The basic idea is to use a public IP address and assign it to the VM to which the access is to be provided by using port forwarding for all ssh requests on port 22. This can be done on the router which is used to connect the server(s) running the private cloud to the Internet. Private cloud is mainly built for accessing the data from the intranet of the organization. The provision is given to the user to access the private data remotely using port forwarding techniques as shown in the Figure 5.5.

This concept was tested on the private cloud using a single public IP address. The server running the private cloud was connected to the Internet using a router. A VM was started on the cloud with a local intranet IP and the router which had the public address assigned to it was configured to forward all requests on port 22 to the VMs local IP address. Now from the Internet, an ssh request was sent to the public IP address.

The request first reaches the router on port 22 as it has the public IP address and not the actual VM. As the router is configured to forward all packets at port 22 to the VM, the request now reaches the VM which handles it and responds back to the IP address from which the request was generated. For providing access to multiple VMs, multiple public addresses can be used and the router has to be configured suitably. Apart from this, the private cloud was used to host a simple website as well by using port forwarding on port 80.

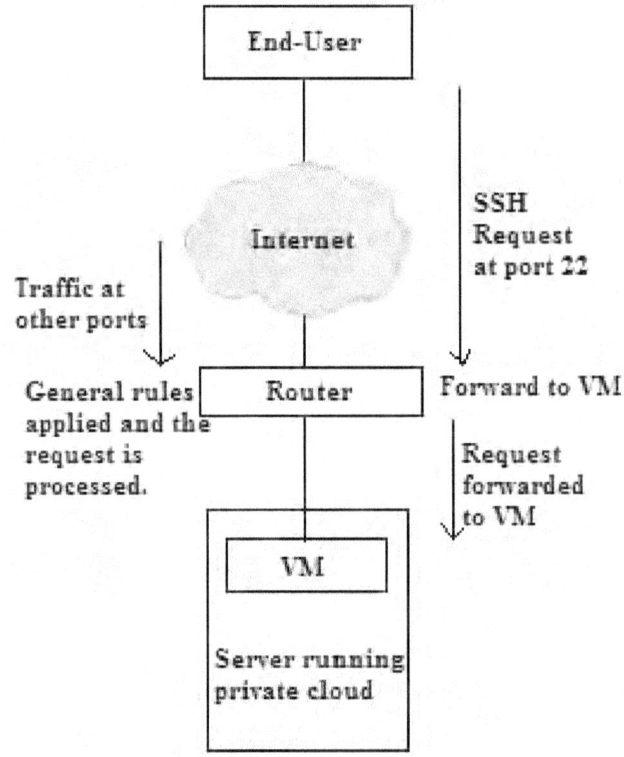

Figure 5.5: Access the private data remotely using port forwarding techniques

Private cloud is the most popular cloud deployment scheme when considering security of data, complete control over the cloud infrastructure, network latency issues and full access of the cloud environment. Many organizations prefer storing mission-critical data in their own infrastructure. Addition of dedicated components is possible when the cloud is owned by the organization. Local Area Networks are much faster than the Internet, thus accessing a private cloud is much faster. Since the cloud is in the organization's infrastructure, it can manage and tweak the services as per the requirements which may not be available elsewhere.

5.4 Experimental Setup and Results

Eucalyptus uses a Linux-based software architecture. One of its many applications is to implement scalable private and hybrid clouds within the enterprise's existing IT infrastructure. Eucalyptus is used to build private and hybrid clouds that provide Infrastructure-as-a-Service model of service. Eucalyptus enables using the enterprise's infrastructure such as hardware, storage, and network with the help of an in-built interface. The Eucalyptus cloud is deployed across the enterprise's on-premise data center which can then be accessed by users using the local intranet. This protects sensitive data against external intrusion attacks thus ensuring its security behind the enterprises firewall. Scalability is achieved in Eucalyptus due to its widely distributed architecture. It contains three levels: Cloud, Cluster and Node.

The Cloud level consists of two components: Cloud Controller and Walrus. Cloud Controller is responsible for providing a web interface for external users and handling all the incoming requests. Apart from that it is also responsible for performing other higher level functions: Authentication, Accounting, Reporting, and Quota Management. The incoming requests from command-line interfaces (e.g. euca2ools) and GUI-based interfaces (e.g. management console of Eucalyptus) is handled by the CLC. Walrus is Eucalyptus storage implementation similar to AWS S3. This provides the ability to deploy additional storage on the available resources. The cluster level consists of two components: Cluster Controller and Storage Controller. A Cluster in Eucalyptus is based on the availability zone concept of AWS. A Cluster is a region in which a particular instance or VM is accessible. The Cluster Controller is responsible for managing the cluster i.e., managing instances and SLAs for that cluster. A Eucalyptus cloud can have multiple clusters. The main responsibility of Storage Controller is to manage the volumes and snapshots for the instances in the cluster. For storing data outside the cluster, back-end storage is used which is available to any instance within that cloud. An additional component may be used for providing an interface to VMware environments called VMware Broker. The Node level consists of a single component: Node Controller. This is installed in the machines where the instances are intended to run. This is responsible for hosting the virtual machines and managing the virtual network. Node Controller consists of a hypervisor (KVM or Xen) which is of hosted type in case of cloud-in-a-box model.

This hypervisor which is installed on the host OS is responsible for hosting the virtual machines. For this purpose, the hardware on which the hypervisor runs must support virtual technology.

Eucalyptus can be deployed in different manners based on the requirements. The types include Faststart or cloud-in-a-box installation, Normal Installation and Eucalyptus HA Installation. The first method of installation is the easiest and fastest method of deployment. This method includes installing all the components in a single machine. This method is very useful and sufficient for understanding the working of private cloud and experiential purpose. However, the other methods do not differ significantly in installation and are pretty much similar to Faststart Installation. The second method involves installation of components in different machines and the last method is useful for large organizations for handling larger loads (High Availability).

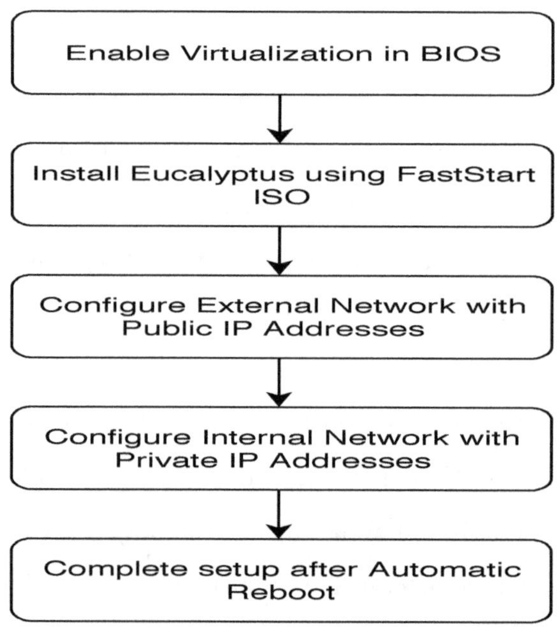

Figure 5.6: Installation steps for Eucalyptus cloud infrastructure

Figure 5.6 gives an overview of Installation of EUCALYPTUS on CentOS. The basic requirement is that the machine(s) involved in the installation of node controller must support virtualization. This is basically because the virtual machines or instances, which require virtualization, run only on node controllers and the other components are used only to manage them. Many machines which support virtualization do not have this feature enabled by default. So it is necessary to enable the virtualization in order for the hypervisor to work. This can be done in the BIOS settings of the machine.

5.4.1 Launching an Instance

An instance can be launched from dashboard home or images tab or instances tab. Creating and launching an instance involves four steps:

- Choosing an image to be used in the instance. The default CentOS image is chosen as shown in Figure 5.7
- Choosing the type of instance. This involves choosing number of instances to be started, instance name and size, availability zone (region of availability of instance) and instance tags (optional).
- Choosing the security for instance. This involves choosing a keypair and security group. This keypair has to be used to access the instance later. A new key pair may be created in order to access the instance from outside the cloud host as these keys are not downloadable from the dashboard.
- Choosing advanced options. It provides feature to add a kernel and RAM disk using their IDs to a custom image. The default image does not need any modification in this matter as all of them are already bundled together. However, when using the default image, user may run into issues regarding the user privileges while accessing the instance. So user uses the following commands in the user data column to create own passwords as shown in the Figure 5.8

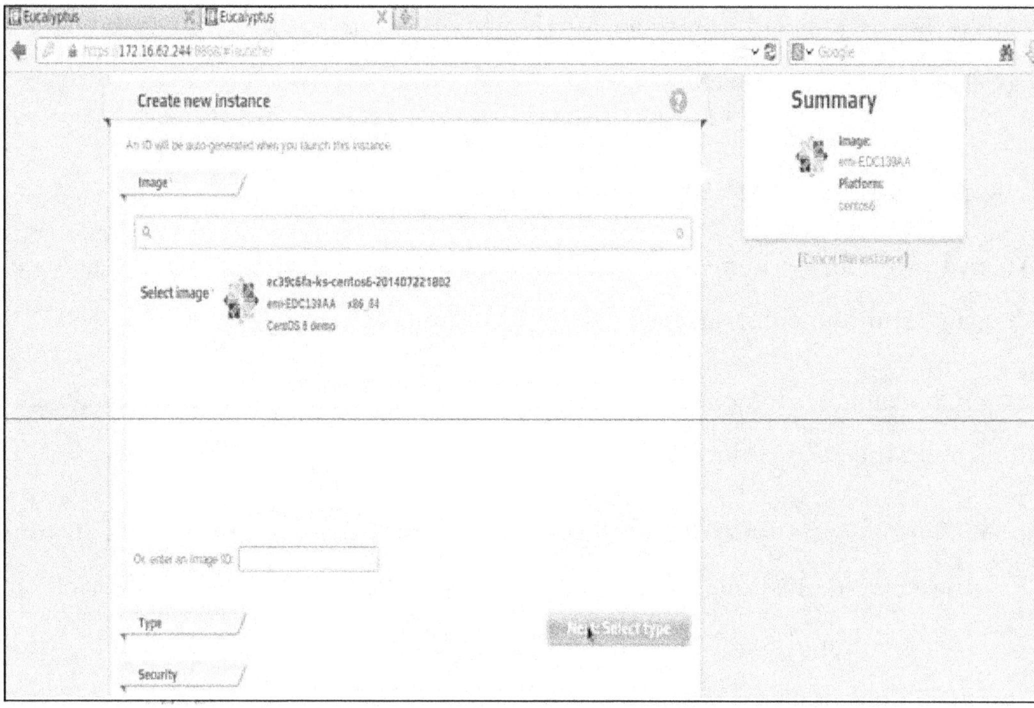

Figure 5.7: Centos image for the selection

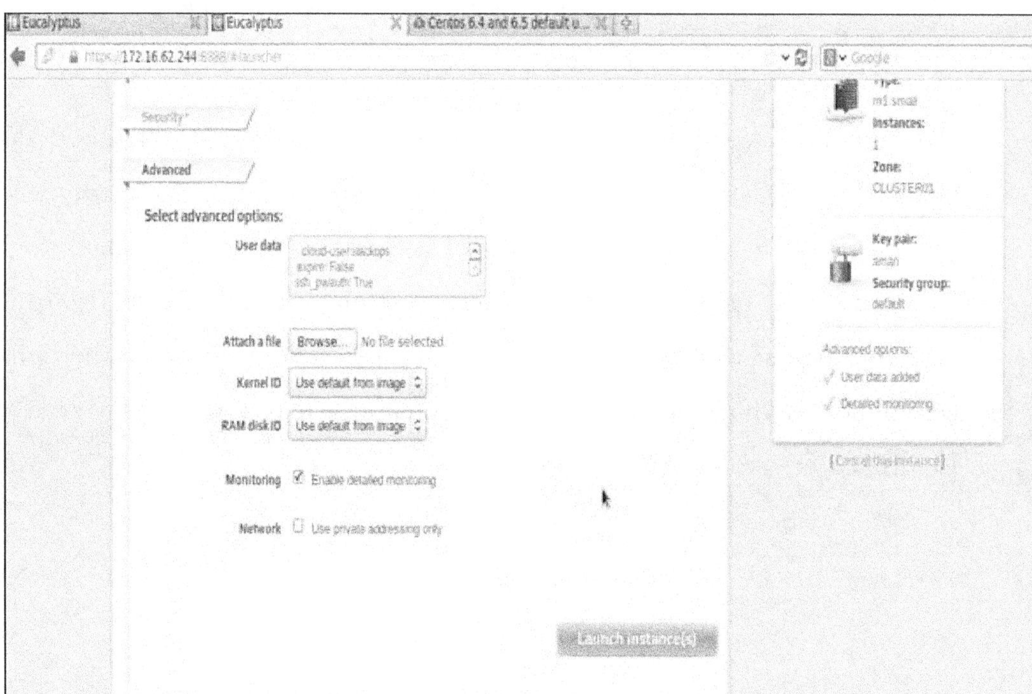

Figure 5.8: Steps to launch an instance

When the instance is launched, a popup should be displayed saying started launch operation for instance i-xxxxxxxx. If some error occurs then it may be due to two

possible conditions. One is that the node controller does not have the requested resources. In this case lower instance size should be used. If not, then the node controller was not registered at the cloud host. This can be verified by using the euca2ools command:
euca-describe-availability-zones verbose
This command should describe all the available resources in the availability zones as described by Table 5.1 which shows the VM Types with their resource requirements in the zone. If all the resources are shown as 0, then there are no nodes registered. However, this is very unlikely in case of cloud-in-a-box installation. Alternatively, the euca2ools command euca-describe-nodes can be used to see all the registered nodes.

Table 5.1: Types of VMs, Their size, and availability

VM Types	CPUs	Ram(MB)	Disk(GB)
Instance m1.small	1	256	5
Instance t1.micro	1	256	5
Instance m1.medium	1	512	10
Instance c1.medium	2	512	10
Instance m1.large	2	512	10
Instance m1.xlarge	2	1024	10
Instance c1.xlarge	2	2048	10
Instance M2.xlarge	2	2048	10
Instance M3.xlarge	4	2048	15
Instance M2.2xlarge	2	4096	30
Instance M3.2xlarge	4	4096	30
Instance cc1.4xlarge	8	3072	60
Instance m2.4xlarge	8	4096	60
Instance hi1.4xlarge	8	6144	120
Instance cc2.8xlarge	16	6144	120
Instance cg1.4xlarge	16	12288	200
Instance cr1.8xlarge	16	16384	240
Instance hs1.8xlarge	48	119808	24000

As shown in Figure 5.9, the instance should now be in a running state and should have its public and private IP addresses. The public IP address assigned to the instance has to be used for accessing it rather than the private address which will not be accessible outside the cloud host.

Figure 5.9: Instance in Running State

5.4.2 Accessing the Cloud

Eucalyptus provides two consoles: user and administrator. These consoles can be used for managing the cloud. These can be easily accessed using the shortcuts provided on the desktop. These are hosted at the cloud host and can be accessed at the cloud hosts IP address using port number 8888 for the user dashboard and port number 8443 for the admin console. For example, if cloud hosts IP address is 172.16.62.244, then the user console can be accessed at https://172.16.62.244:8888/ or simply https://localhost:8888/ from the cloud host machine and the user console can be accessed at https://172.16.62.244:8443/ or simply https://localhost:8443/ from the cloud host machine. If this does not work, then either the eucalyptus service has not started properly on the cloud host or the cloud host is inaccessible from the requesting hosts network. This opens up the dashboard and credentials are necessary to proceed further. The credentials should be account name: demo, username: admin, Password: password for the first login. However, it is advisable that these are changed later for obvious reasons.

This should give us the access to the user dashboard. The dashboard home shows the number of all the current running instances, stopped instances and scaling groups in different availability groups. It also shows the status of storage volumes and snapshots along with number of security groups, key pairs and IP addresses assigned/available. Thus it gives us an complete overview of the cloud with respect to that user. An image is an operating system file that is used to launch an operating system in the instance when it starts. By default, a centos image is available for use in the dashboard. Images can be viewed from the images tab.

Summary

The chapter presents and demonstrates the deployment of the developed system in the cloud based environment. The creation of public and private cloud setup and deployment techniques are discussed. The subsequent chapter discusses the real time capturing and sharing of routine health data.

Chapter 6

Validation of the Proposed Electronic Health Record System

The developed EHR system is tested and validated with realistic data which is described in this chapter.

6.1 EHR System Testing

Healthcare IT sector in India is expected to reach $280.25 billion by 2021, with an annual growth rate of 15.9% [109]. Especially EHR system needs to satisfy quality requirements to provide better care to the patient. To achieve this software testing is necessary for every healthcare industry. The testing is needed for EHR system to validate the functionalities, interoperability, compliance and performance. During the software development life cycle, application developer should give suitable time and testing effort for the developing system. If not, there is a possibility that issues could have been found early at the implementation.

Web application consists of numerous application entries and exit paths, different type of browser with end users, people with varying technical skills using software which makes testing application a challenging task. However in this software each module is tested individually with some input. All the requirement were preserved and taken care while testing each module individuality [110]. As it is web application, unit testing, functional testing, system testing, Interface testing and security testing needs to be conducted for the EHR Systems.

6.1.1 Unit and Functional testing

Unit testing can be conducted on individual units of the software units or a group of related units. Mainly in web based application, it is used as a verification and validation method. As a result, developer gains confidence that individual module of the application source code are appropriate to use further[111]. A manual approach mainly focuses on testing the links of database connection, verifying the submitted details from the user to forms and web pages. The web application requires high data consistency where the data integrity and error check helps in performing edit, delete or modification of forms or do any DB related functionality. The unit testing and functional testing details are presented in Table 6.1

Table 6.1: Unit and Functional testing

Test	Component	Responsibility	Accept/Partial
ID-01	Each major function performs as specified in the report.	Login -Using the verification method to ensure that authorized users get a login and deny other user -Login with invalid credentials negative test - Verify Username and Password is encrypted and not stored in plain text anywhere Hospital Administrator: -Register, reinstate and terminate a patient -Alerts to patients in case of missing dependencies -Remove and add a dependent	Accept
ID-02	Details related to design changes and customizations needs to document for future reference.	-Drop down, click etc work properly. -Navigation works properly -Tested with partial data and exits the browser -Display of charts to show patient past record etc	Accept
ID-03	Test the appearance of the screen as user expected including web content, position of fields etc.	-Tried with different browsers -Entered invalid data types and check the error message -Entered future dates and check for the error message -Alerts to patients in case of missing dependencies	Partial
ID-04	Test for spelling errors,color changes and readable icons.	The system is reviewed by a healthcare expert who has not used the application before	Accept
ID-05	Test for web page entries that have been corrected and later corrections are displayed accurately on the page.	Made changes in the existing records and signed out from the system. Later logged in to the system again and check if data there present.	Accept

6.1.2 System Testing

The system test verifies all system functionalities and see the over all system functions achieved or not. Developer level testing has to be done for each process before exposing the feature to the actual prototype. System testing compliance with its specific requirement are shown in Table 6.2.

Table 6.2: Table of System Testing

Test	Component	Responsibility	Accept/Reject
ID-01	Test the flow of data between different modules of EHR System.	-Perform EHR workflow. Example : 1. Register Patient 2. See eligible information 3. Register all healthcare professionals 4. View details 5. Edit details 6. Exit -Manage patient flow from check-in through check-out.	Accept
ID-02	Test the interface by sending and receiving data completely and accurately in both the direction (Unidirectional)and (Bidirectional).	- Sending and Receiving healthcare data between healthcare providers and patients are tested. - Moving patient data between different levels are tested	Accept
ID-03	Test the connectivity and instance access of data with other healthcare facility is accurate and authorized.	-Patient health record transfer functionality enabled if patient need clinical continuity in other hospital	Accept
ID-04	Test the access privilege of the EHR system by conducting test attempts to gain access when not authorized.	Log in as a user and try admin actions like deleting updating or terminating patient record membership	Accept
ID-05	Data recorded in tables, graphs patient summaries, health reports, etc are accurate.	Database consistency check	Accept
ID-06	Tested for the data visibility of the forms while displaying in the web documents.	-Portable records are accurate -Patient history is maintained and accurate -Database consistency check	Accepted

6.1.3 Compatibility Testing

Compatibility testing is very important testing aspect for cloud based web applications. For e.g. Browser compatibility: Different browser has different setting and ways of rendering a pages. Following browsers are used for testing

- Internet explorer
- Microsoft edge
- Google chrome
- Opera 10
- Mozilla Firefox

There was minor changes in the UI with thick border at certain places and color with different browsers. However all the features of application were equally accessible from all the browsers.

6.1.4 Security Testing

EHR system contains the sensitive patient information and the data should be systematically tested. HIPAA has defined security rules to safeguard the EHR system, it defines defensive features to fight against data security violations [112]. All the activity of the users in the EHR system must be recorded to validate the activity log. In the cloud based EHR system, plot the various dynamic graphs for all users which showed the frequency of different activity. Furthermore, hospital administrator could keep a track of various features like percentage of different users, number of new registration etc. Every patient can see different activity they were perfuming in the web application and analyze themselves. The information available is a highly individualistic and is available only to respective users. Same as applied to all kind of healthcare users on the web application. The Algorithm 6 describes the activity tracking technique used in the EHR system.

Algorithm 6 Activity tracking

```
Begin
plist=session['plist']
plt.figure(fgsize=(16,8))
plt.title('Track your activities')
plt.xlabel('Activities you did')
plt.ylabel('Frequency Of Activity')
plt.xticks(fontsize=8)
vl=['Login','Logout','Viewed Prescription','Viewed Bill','Patient Record','Lab
    Record','Transfer Request']
plot=plt.plot(vl,plist,'-o')
img = BytesIO()
plt.savefig(img)
img.seek(0)
response = make response(img.read())
response.mimetype = 'image/png'
End
```

Plist in first step is the list of statistics calculated for each activity performed in the EHR system. Pyplot module of matplotlib is used to plot a graph of different activity in the system with data plotting method. Make response function is used to convert the image of plot into a response that can be returned and rendered on an HTML template. Following are the some of the test done as shown in Table 6.3.

Table 6.3: Security Testing

Test Description	Output Message
Test by passing output URL directly into browser address bar without login	Access denied
URL injection by manipulating query string parameters of second page	Access denied
Login with invalid user name and password	User name and password did not match

6.2 Validation of EHR System Using Realistic Data

To validate the capturing and sharing of the developed EHR, the PHC, SHC, and THC healthcare Facility in Udupi district, Karnataka was visited. The permission has been obtained from the District Health Officer(DHO) to capture anonymised patient data and test the system. The process followed in this research is in no way compromise privacy of the patient. To validate the EHR system with realistic data, about 100 patient records including 60 from PHC and 40 from Community Healthcare Center are entered into the EHR system with the assistance of healthcare professionals. The confidentiality of the patient health data has been ensured while entering the data into the system. The lab and pharmacy data were entered using the prescription written by the healthcare professionals at the respective healthcare facility.

To validate developed EHR system further with Secondary and Tertiary level, the realistic data, about 30 patient records including 20 from District hospital, Madikeri(Secondary Level) and 10 from Kasturba Hospital, Manipal(Tertiary level) are entered into the EHR system with the assistance of healthcare professionals. Using this digital form of data, patient record can be accessed at any place, any time and at any hospital.

When the EHR is tested in the real-time environment, the result indicated no errors while capturing and generation of patient health information. In addition to this, it also evidenced the availability and generation of the vital statistical report about the patients visit to the PHCs. This arrangement allowed the healthcare professionals of other levels of the healthcare system to view the patient data and provide quality healthcare to those who reported to them for receiving care. The EHR data stored in the system is made accessible to the healthcare users through secure login. It is also verified that the patient who needs special treatment, the organized health reports of the patients are available for instant sharing with the referring doctor using the cloud-based system.

The system also allowed healthcare professionals to generate various statistical reports. These reports are presented in terms of the graph that allow the professional to have a better view and assist them in taking quality decision and action. The sample graphical representation of collected data from PHC and Community Health Centre is depicted in Figure 6.1 and 6.2 respectively.

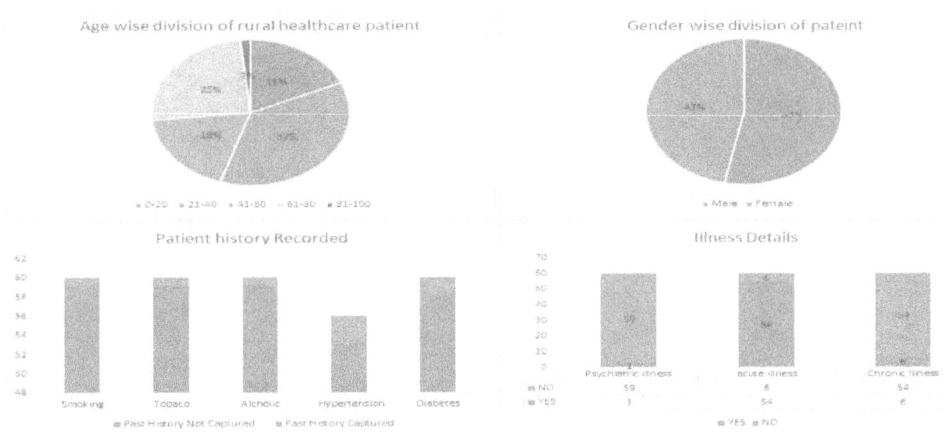

Figure 6.1: Basic demographic details and illness status of patient in primary healthcare centre

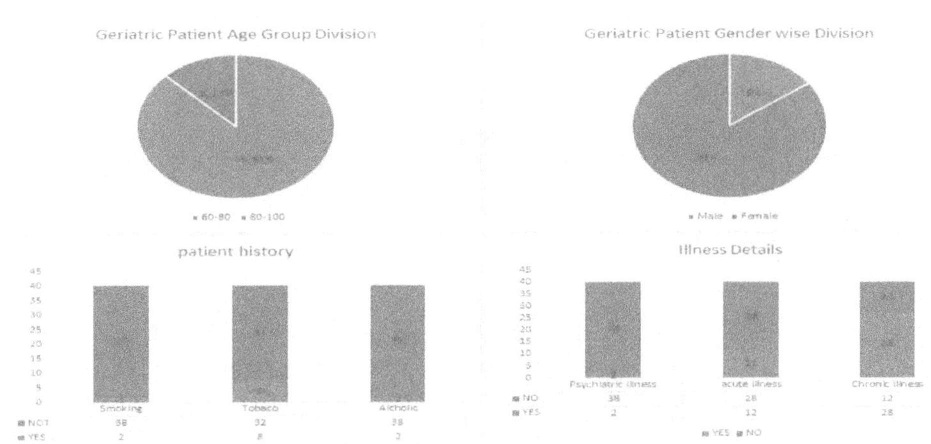

Figure 6.2: Basic demographic details and illness status of the patient in Community healthcare center

The system also allows the doctors to generate the patient health summary and BMI (Body Mass Index) graph to view the progress of care provided and take necessary action in case if the patient reflects any unexpected condition. The sample graphical view of BMI report and patient health summary sheet is depicted as Figure 6.3 and 6.4, respectively.

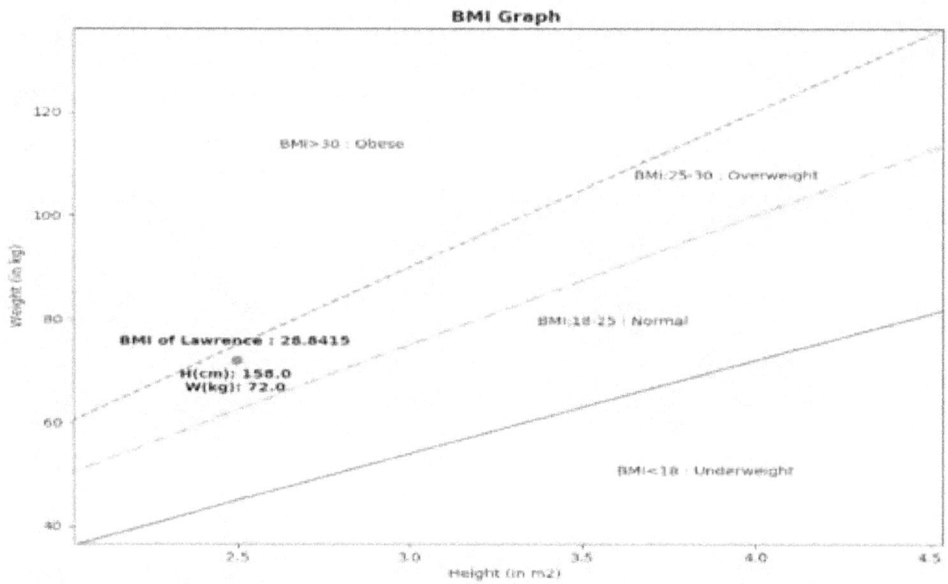

Figure 6.3: Graphical View of BMI

Discharge Summary

Patient Name:
Patient Aadhar Number:
Admission Date: 2018-12-07 07:01:53.147631
Discharge Date: 2018-12-08 07:31:39.878685
Attending Physician:
Dictated By:

Condition on Discharge Stable
Final Diagnosis Pneumonia
Procedures None
History of Past Illness Admitted on with shortness of breath ...
Discharge Medication , ,
Discharge Instructions Follow up after 3 weeks
Primary care Nurse

Figure 6.4: Sample format of patient health summary

The EHR in this study is developed with an understanding of the patient data flow among the three levels of the healthcare system (Primary, Secondary, and Tertiary).

The EHR supports health record transfer, when a patient moves from one healthcare to other level. The patient health summary is made available to any level at any time. This arrangement allowed the healthcare professionals of other levels of the healthcare system to view the patient data and provide quality healthcare to those who reported to them for receiving care. The Role-Based Access Control (RBAC) of the system allowed the referral level doctor to view the patient data without any difficulty with valid authentication. Whenever referral level doctor needs more information about the patient, referral hospital administration can send request to referred hospital after taking consent from the patient. Once the doctor receives authorizations, he will have access to patient data. This record transfer approach allows the doctors at the referral level to get access to the complete patient demographic data, patient history, diagnosis information (CT scanners, MRI scanners) and pharmaceutical information which is stored in the EHR database. Here, the referral doctor can view the patient data and if required, update the diagnostic and therapeutic details of the patient. The analysis of the recorded data of EHR also reveals that the patient who needs special treatment, the structured patient reports are available for the immediate sharing with the referring doctor using the cloud-based system.

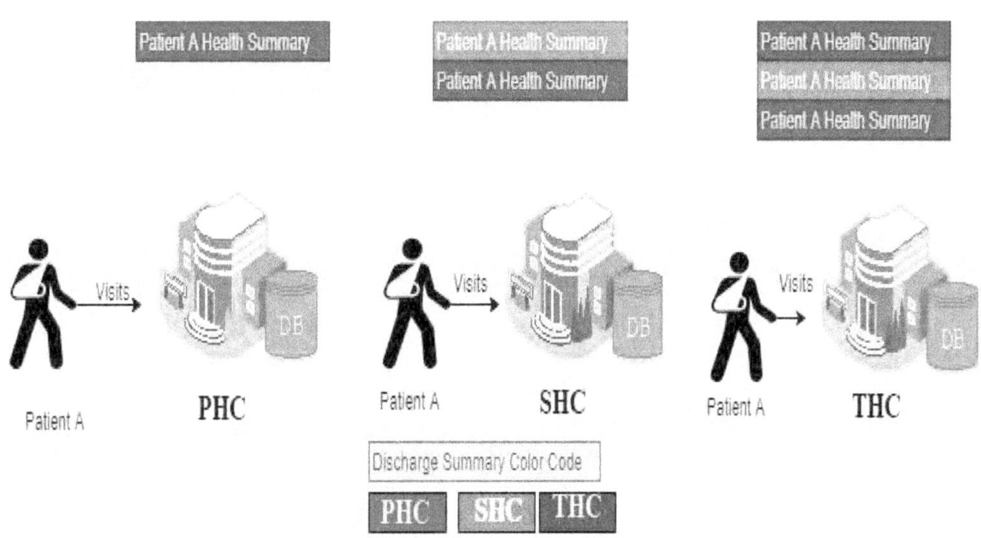

Figure 6.5: Patient record transfer from one healthcare level to other

Figure 6.5 depicts the complete sharing of patient data from primary level healthcare facility [as depicted as Hospital A in the Figure 6.5] to secondary and tertiary level healthcare facility [as depicted as Hospital B and C in the Figure 6.5].

At every healthcare level, the doctors authenticate the access using the digital signature. After writing prescriptions, doctors received an email which redirected them to a page where they could review what they prescribed and then digitally approve it with their signature. This approved prescription was automatically visible to patient's, in the form of printable PDF as soon as it was approved in real time. This feature enables and enhances the secure capturing and sharing of patient data as the EHR does not allow the doctors to approve any documents without the authorized digital signature as shown in Figure 6.6. During the entire episode of capturing to sharing of patient data, the developed EHR has proven to be successful with no errors or bugs.

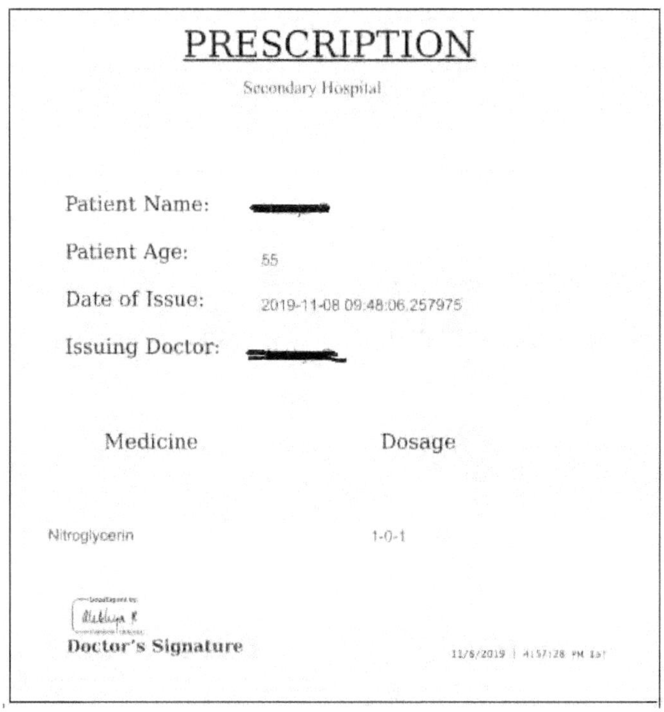

PRESCRIPTION

Secondary Hospital

Patient Name:

Patient Age: 55

Date of Issue: 2019-11-08 09:48:06.257975

Issuing Doctor:

Medicine	Dosage
Nitroglycerin	1-0-1

Doctor's Signature

Figure 6.6: Digitally Signed Prescription Sample

Doctors have the provision for viewing DICOM images with adequate clarity along with all the associated tags. In traditional method, doctor usually had to use a separate software for viewing medical images which is time consuming and consumes more space. Whereas the proposed EHR system is customized solution and can be integrated with any kind of diagnostic and imaging equipment to capture and share in a secure environment. Further, In the developed EHR, DICOM image viewing is tested by transferring the record from one healthcare facility to other. It is observed that referral level doctor can view the DICOM image without any delay. The sample DICOM images with associated patient details are depicted in the Figure 6.7.

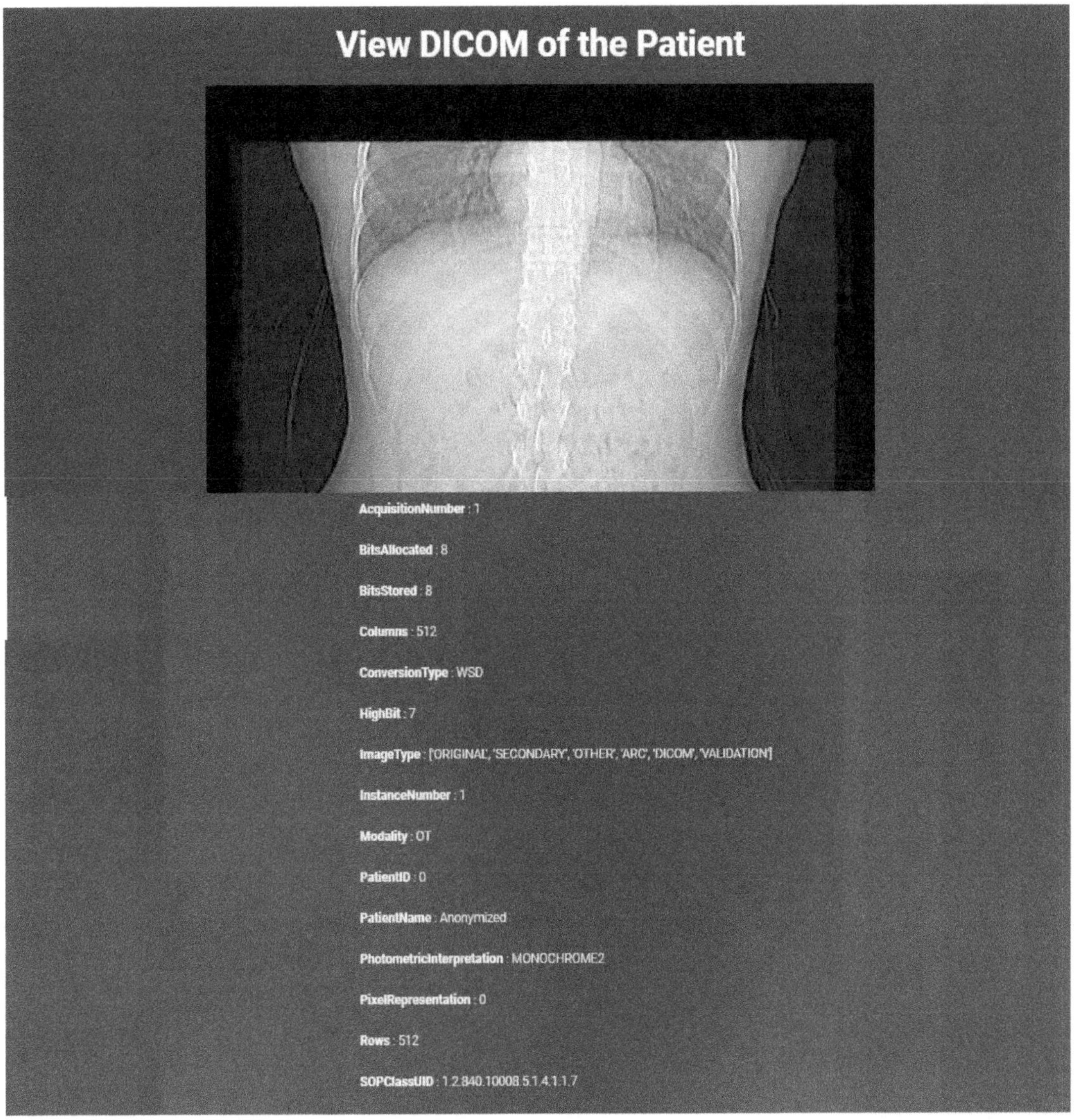

Figure 6.7: DICOM View of the Patient

6.2.1 De-identification process in EHR system

De-identification is the process used to prevent personally identifiable information from being revealed [113]. The de-identified data is mainly used for clinical research, academic research, archiving, statistical analysis, policy formulation, the development and promotion of diagnostic solutions and for facilitating the health. In this model, when health information of the patient needs to released to the other level, the referral hospital administration needs to take consent from the patient. If the patient wishes to hide the identifiable information from his/her record the de-identification procedure is applied on the health record of the patient. During this process certain pieces of information (which gives the identity of the patient) from the record is removed. During this process a new ID is generated for the patient record and the same is used in the other healthcare facility. If the other healthcare facility provider needs the identity of the record, with permission, the data is decoded back to the original record as shown in the Figure 6.8.

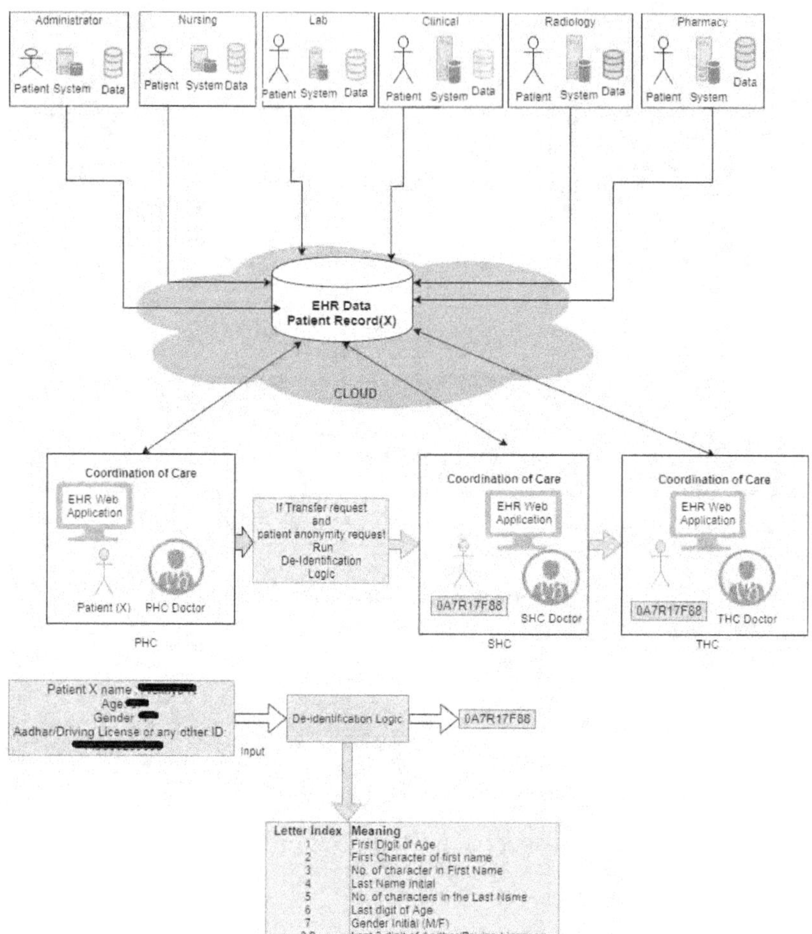

Figure 6.8: De-identification in EHR flow.

6.3 Credibility factor of EHR system

The main responsibility of Healthcare IT is to improve the health of people through innovation research in healthcare sector. A cloud based EHR system assists the various levels of healthcare users to attain best patient outcome, reduce the cost, improve clinical outcome, and save the patient time by reducing the waiting time. Promptness in delivering patient care is considered as one of the most important factor in assessing the quality of service offered and provided by the healthcare setup. To make the cloud based EHR system more powerful, it needs to identify how well system supports the functionalities to deliver quality of the care to the patient. One such solution implemented is credibility factor. The credibility factor score helps healthcare IT team to improve the features of cloud based health information system. The ultimate objective of any healthcare I&CT implementation should be to assist the end user in better decision making and quality of healthcare to the patient.

Table 6.4: Credibility factor positive aspect questions

Sl. No	Parameter	5 (Strongly Agree)	4 (Agree)	3 (Neutral)	2 (Disagree)	1 (Strongly Disagree)
1	EHR system deliver quick and consistent access to patient data	5	4	3	2	1
2	EHR system reduces error	5	4	3	2	1
3	EHR allows easy transfer of data	5	4	3	2	1
4	EHR system provides better graphical view of data	5	4	3	2	1
5	EHR increases trust on doctor	5	4	3	2	1

Table 6.5: Credibility factor negative aspect questions

Sl. No	Parameter	5 (Strongly Agree)	4 (Agree)	3 (Neutral)	2 (Disagree)	1 (Strongly Disagree)
1	The EHR system is too complex and not convenient to use	5	4	3	2	1
2	EHR increases health professionals workloads	5	4	3	2	1
3	EHR requires longer time to enter patient data than paper based	5	4	3	2	1
4	Using an EHR detracts from my clinical effectiveness	5	4	3	2	1
5	EHR system compromises patient safety	5	4	3	2	1

Credibility factor of EHR system is computed using multidimensional aspect parameters. Numerous dimensions that are considered to identify patient satisfaction includes registration, doctor care, nurse, lab facilities, and the pharmacy. The parameter identified are categorized into positive impact (M) and negative impact (N). The main expectation of any level of healthcare professionals is to use the system in the following manner. The implemented system should reduce their time in recording and standard coding and they need to spend more time for patient care and assessment. Keeping this in mind, to rate the EHR system in each user level different set of parameters are identified and depicted in Table 6.4 and 6.5.

Table 6.6: Credibility score table

Credibility Score	Grade
21-25	Excellent
16-20	Good
10-15	Average
< 10	Needs Improvement

The computation of credibility factor is depicted by a flowchart in Figure 6.9. All the healthcare professionals are part of the system and using their user credentials he/she can login and use the EHR system for providing the feedback. To collect feedback from the various healthcare users, the validated and tested questionnaires are used. Visited Statistical Consultancy Services(Dept. of Statistics) in MAHE, Manipal and discussed and identified the parameters which are essential for computing credibility factor and same features are incorporated in EHR application. The questionnaire record the responses in a scale from from 5 to 1(Strongly Agree to Strongly Disagree). Credibility of the EHR system is computed based on the questionnaire responses. Equation used to compute the credibility score is as shown in the equation (6.1)

$$CredibilityFactorScore(CFS) = M - N \tag{6.1}$$

Using the credibility score grades are defined like Excellent, Good, Average and Needs Improvement as shown in Table 6.6. The threshold level is defined in the calculation is 10. In the given table, if the calculated value of the credibility factor is greater then the threshold, the EHR system is considered as stable. If credibility score is less then threshold then EHR system needs an improvement. This feedback could help to further improve the presentation of EHR system for better and more efficient patient care.

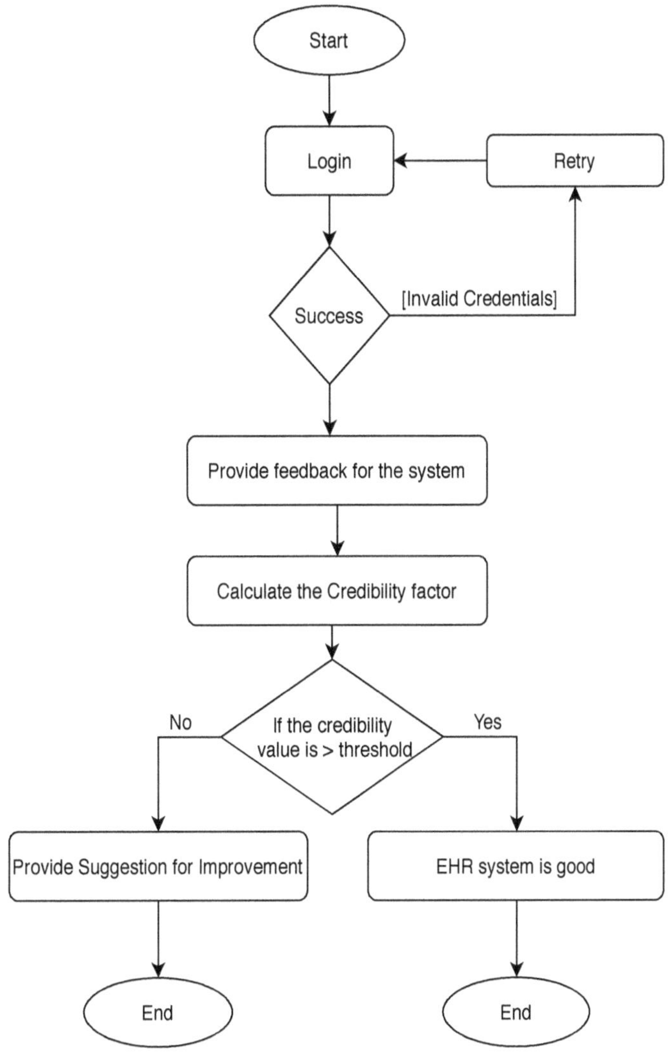

Figure 6.9: Flow digram for computing credibility factor of the EHR system

6.4 Comparison of EHR system with other existing system

The comparison of EHR functionality with other existing systems are discussed in Table 6.7. Developed EHR system is compared with the other systems including Open EMR ,Open EHR ,Open MRS, GUN health and OSEHRA. Northern European countries mainly are using OpenEHR for digital storage. In India, Africa and other south east Asia Open Medical Record System(OpenMRS) is developed but the care is limited to only one hospital system. The records are not able to be exchanged with other health-care levels. With proposed and developed standardized EHR system, records are electronically documented in the system and is available instantly at any location. In the comparison study 'Yes' indicate the focus criteria considered for the EHR system fulfills completely. If the criteria considered for the EHR system does not satisfies completely, the end user requirements is marked as partial. E.g. The transition of care is considered as partial only in the OpenMRS, the reason is the continuity of the care is provided within the same hospitals of different department. If the patient is transferred to other hospital the continuity of the care is broken. None indicates the focused criteria is not at all implemented and not give any kind of service to the healthcare system.

Table 6.7: Comparison of proposed EHR system with other System

Focused Criteria	OpenEMR	OpenEHR	OpenMRS	GNU Health	OSEHRA Vista	Proposed EHR
Transition of care	Yes	None	Partial	None	Yes	Yes
Clinical decision support system	Yes	Partial	Partial	Partial	Yes	Yes
e-prescribing	Yes	Yes	None	Yes	None	Yes
Patient specific resources	Yes	None	None	None	Yes	Yes
Clinical summaries	Yes	Partial	Yes	None	Yes	Yes
Authentication, Access control, Authorization	Yes	Yes	Yes	Yes	Yes	Yes
Emergency Access	Yes	None	Yes	None	Yes	Partial
Integrity	Yes	Yes	Yes	Yes	Yes	Yes
Automatic log off	Yes	None	None	None	Yes	Partial
Incorporation of laboratory test and result	Yes	Yes	Partial	Partial	Yes	Yes

The comparison result shows that all the criteria works completely except the two criteria including emergency access and automatic log off. The main reason is that, in the proposed system emergency access works only with patient unique identity and also it needs some special algorithm like break the glass approach during emergency situation. In case of emergency situation the system has two challenges to address those are:

- The patient has arrived at the hospital due to an emergency. The patient is conscious but is not able to give all the correct information that is required to authenticate. The patient doesn't want all his/her data to be visible to a new hospital. Hence, he/she tells the emergency details required for an emergency login. This enables the staff to log in and get the data that is crucial for the first aid as well as maintaining the privacy of the patient in the new hospital.

- The patient has been bought in by the emergency response team who had met with an accident. He/she is in an unconscious state. The staff tries to gather any information about the patient's name or his/her hospital id number. In case, they fail to gather the above information, they try to gather the visible information that is required for emergency login. This includes collecting the height, weight, birthmarks, iris color and other information which is used as a filter to make the search easier and faster. Once the patient details are entered in, the search results appear. It shows the basic required information that can be used to treat the patient.

The automatic log off is the process of logging out of the system automatically when healthcare user forgot to perform the option of log off. The system goes into the inactive state for some period of time and again resume back after providing credentials for the successful login. Most of the system, healthcare users deactivate the automatic log off option because, they must re-enter the log in details again. In the proposed EHR system,the automatic log off criteria showed as partial because patient examination time varies from patient to another. During this period continuous access of the patient is very much essential.

The designed EHR system has the following performance benefits :

- Interoperability: The proposed system has provision to integrate any biomedical, diagnostic, and imaging equipment to capture and share patient data. Support is provided for the delivery of effective patient care by capturing and transmitting the electronic data from one healthcare facility to other.

- Ease of access: The patient record can be accessed from anywhere at any time when healthcare providers are looking for patient data.

- Improve patient data safety: The data which is available in the EHR system is secure and digital representation of record reduces the theft of the patient record.

- Paper documentation: The proposed system reduces 75% of the paper documentation.

- Scalability: Provision is made in the proposed system to extend the services and capacity when multiple healthcare levels data are added into the EHR system.

6.5 Case Study on EHR system

In India, healthcare services are mainly delivered through three levels of healthcare systems including PHC, SHC and THC. EHR has potential to make available the real time data of the patient to the healthcare professionals by connecting all levels of public healthcare system, irrespective of their geographical boundaries. The case study gives idea about how healthcare professionals from different care facility access the data of the same patient in different levels of healthcare facility. The EHR system is capable of generating the reports of the patient's visit to the different levels of healthcare facility. This arrangement allows healthcare professionals of other healthcare facility to view the patient data and provide quality healthcare to those who reported to them for receiving care. In this case study it is assumed that the patient is from rural place and for his/her treatment visits PHC level first. Patient named P1, started experiencing chest pain and palpitations and so visited the PHC on 11th September 2018. Once registered to the EHR system, patient moves to the nurse, where the basic vital signs such as height, weight, blood pressure are collected and recorded in the EHR system. On completion, patient consults the doctor where doctor takes the history of the patient and physically examines the patient. Once the doctor confirms the diagnosis, recommends the prescription to the patient. The treatment details of the patient are entered using ICD 10 code R07.1 for chest pain. The prescription details are entered suing NDC code as analgesic medicine [54162-555-01]. All the details including registration of the patient, vital sign details, diagnosis and medication details are stored in the patient health record in the EHR system. The PHC deals with the patients whose medical conditions can be managed on an outpatient basis. So, the patient is referred to SHC for further treatment. The health status of the patient at PHC is depicted in Figure 6.10.

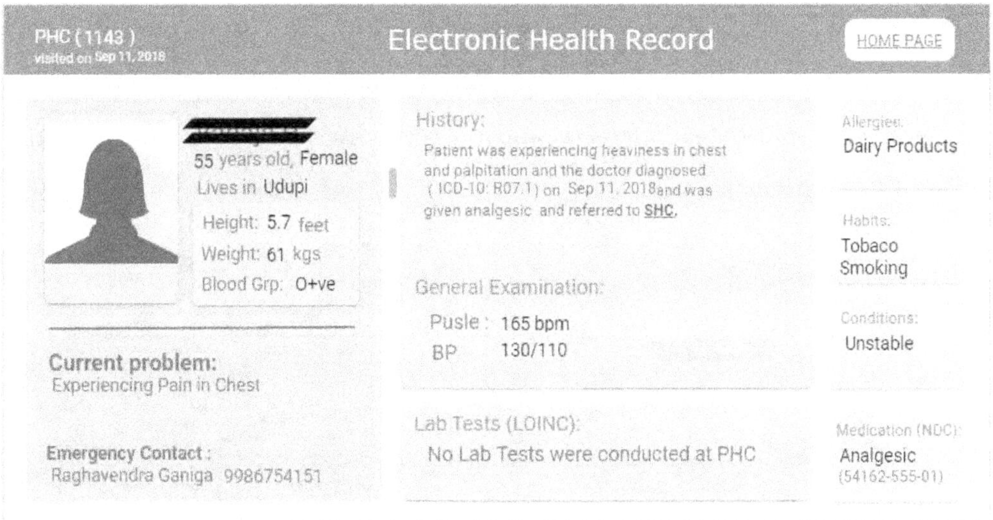

Figure 6.10: Patient health status at PHC

As per the PHC doctor's recommendation, the patient visits the SHC on 13th September 2018. When patient gets referred from primary to secondary or tertiary level for specialized treatment, the health care professionals at the referral level will have access to patient health summary sheet. Doctor at SHC after conducting general examination, suggests the patient to undergo ECG (Tropical T Test) LOINC:33204-9. The patient is diagnosed with myocardial infraction ICD-10:I21.9 and is referred to THC for advanced investigation. All the patient health information details are updated in the EHR system. The health status of the patient at SHC is depicted in Figure 6.11.

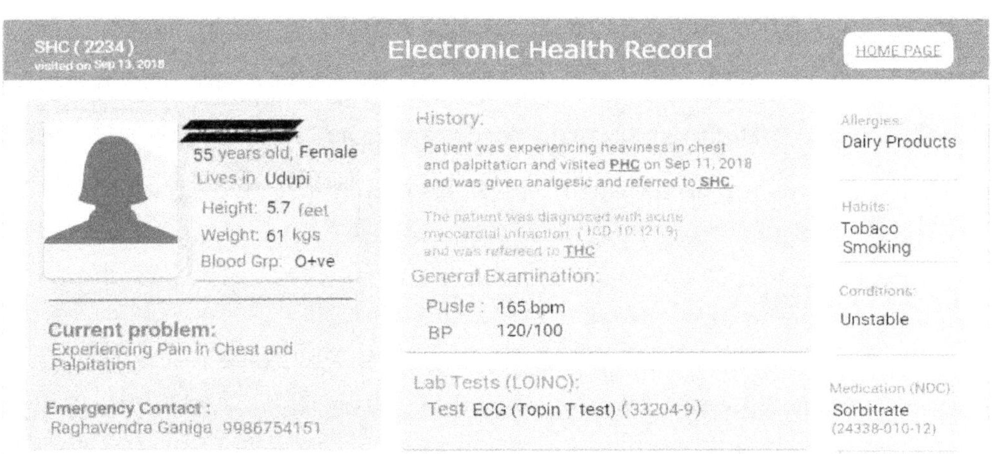

Figure 6.11: Patient health Status at SHC

Patient visits the THC on 27th September 2018. Doctors from THC requests to view the complete record of the patient. An instant transfer request is sent to the level from where the patient is refereed, and on receiving authorization, he/she will have access to the patient record.

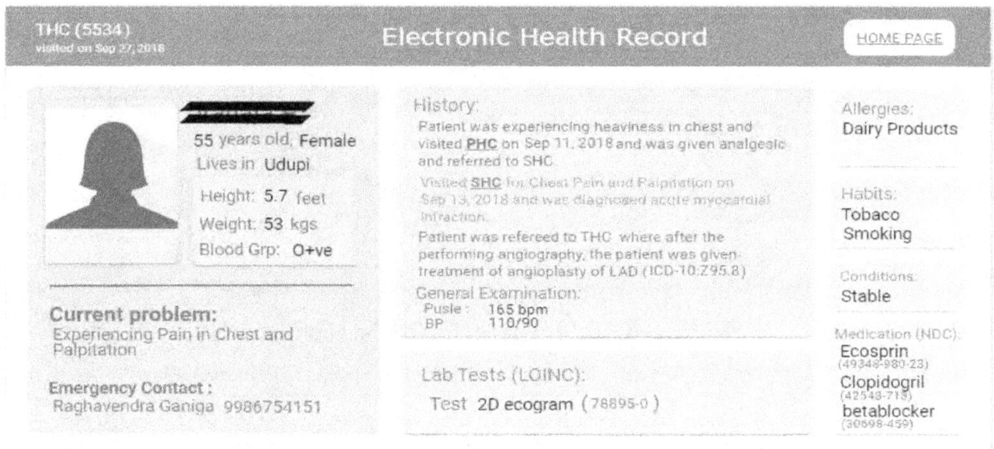

Figure 6.12: Patient health Status at THC

The health status of the patient at THC is depicted in Figure 6.12. Now doctor from THCs get access to the complete patient demographic, history, diagnostic, and pharmaceutical information. Here, referral doctor can update the diagnostic and therapeutic details of the patient. Doctor suggests to undergo 2D ecogram (angiography) LOINC:78895-0 and after analyzing the lab results, the patient is recommended the treatment of angioplasty of LAD [ICD-10-CM Z95.8]. The patient is given medications (Ecosprin[49348-980-23], Clopidogril [42543-713], betablocker[30698-459]) for next 3 months. During the entire episode of capturing and sharing of patient data, standards such as ICD-10,SNOMED-CT,LOINC, DICOM standards are used, which represents the case study of patient P1.

Summary

The chapter presents and demonstrates the validation of EHR system with realistic patient data of primary, secondary, and the tertiary care levels. The subsequent chapter discusses conclusion and future scope of the research.

Chapter 7

Conclusion and Future Scope

The aim of this research has been to design and develop a cloud based standardized EHR for public healthcare facility and provide a solution to access patient health information by enabling healthcare data standards.

7.1 Conclusion

Indian healthcare scenario follows three tier structure including PHC, SHC and THC for delivering healthcare services to the population. Healthcare system is diverse, consisting of a large number of hospitals of various size and units, run by the central and state government. Traditionally in India, patient health records gets documented on paper based format. Capturing and sharing of such information becomes a challenge for the healthcare professionals and providers. Paperwork has always been a hassle, not only in offices and institutions but hospitals as well. Here, the patients are required to carry their medical reports to every hospital they visit. This task is not only painstaking but can also lead to unnecessary test or wrong diagnosis due to absence of the same. Presently some of the healthcare facility started using Information Technology(IT) in healthcare to manage patient data in digital format. Due to lack of standardization of the record, health information of the patient available in the some healthcare data repository cannot be accessed by the others. Healthcare providers always needs the patient health information at their fingertip for providing timely care to the patient.

Considering the above, the objective of this research was to provide a simplified solution to completely support the healthcare service providers of PHC, SHC and THC with complete, accurate, adequate and timely patient information for providing quality care during the episode of care. Advancement in IT technology in healthcare sector has made it possible to maintain and manage the patient data in digital form in all levels of healthcare system. With the Government of India backing the Digital India movement, there have been a lot of advancements in the conglomerations of Healthcare and IT industry. This lead to the vast developments in the methods of record maintenance for the Indian healthcare sector. This is where EHR came into picture. EHR is a digitized version of a patient's health records including administration, nursing, Laboratory, clinical, Radiology, and pharmacy modules. It is a trump card in Healthcare domain as it provides secure real-time information retrieval. Now a days the web based EHR system is beneficial to store the patient's health information. This data should be globally available across various health care levels.

In this work, a cloud based standard EHR system is developed to manage patient and community health data to improve public health surveillance and assist in controlling communicable and chronic diseases. Various healthcare standards coding for recording and reporting such as ICD 10, SNOMED-CT, LOINC, DICOM and NDC has been incorporated. The system has better interoperability, maintainability, and scalability compared to traditional paper based health information system. The healthcare standard and medical coding with cloud database enable connectivity between PHC, SHC, and THC levels to make patient health data available all-time when required for his/her care.

Two cloud deployment models namely public and private have been used to manage patient health information in EHR system. The public cloud infrastructure uses Amazon Light-sail services with high availability. This model is robust and is likely to operate uninterruptedly without failure for a long time. Other model used especially for the rural healthcare system is the private cloud deployment model.

Health Information System contains sensitive information about the patient and the data needs to be protected from the intruders to safeguard the information. To share the data among multiple healthcare providers in a secure manner, EHR reference architecture is developed and various attacks and vulnerabilities have been identified. Potential threats are identified in the EHR system by using STRIDE model and the amount of risk posed to the systems are calculated using DREAD model. A new framework for highly secure sharing of EHR in cloud environment is developed which mainly focuses on Authentication, Authorization and Attribute based encryption. SAML-SSO authentication model is adopted in EHR module to access multiple services of health information after signing in only once which improves the security and increases the access speed of the health information service. For Authorization Oauth2.0 is used which gives resource access in the cloud based environment. ABE eases the process of managing key for the encryption and it makes it possible to access the patient health information based on the professional roles, experience, and specialization of the doctor and health-care levels. Various modules for EHR has been developed for Indian health-care scenario. Identified security features of EHR ensures a lowest risk of breach and enhanced control over the system.

To validate the developed EHR system in real time, the healthcare facility in Udupi district , Karnataka was visited. The developed cloud based EHR system has been validated with real patient data of 60 from PHC and 40 form CHC. The test result indicated error-free capturing and sharing of patient data at various levels of the healthcare system. The present EHR includes the minimum data sets suggested by MoHFW, India. These features of EHR system would allow the planner and manager at all levels of the public health system to plan and conduct various preventive, curative and rehabilitative programs and ultimately support them in achieving national and international goals set for public healthcare.

7.2 Scope for Future Work

Research always come across a number of open questions that allows to look forward and pursue the course of action in making the research outcome more feasible and acceptable. Some more advanced features such as interface to digital writing technology, audio recording during consultation, and integration of disease related awareness in local language could be built for EHR system.

One additional feature that can included in the EHR is to capture patient real time data, when the patient is moving from his/her home or incident site through ambulance or any other mode of hospital transport. In most cases, hospital fails to capture these data which is vital in advising the paramedical and emergency technologist to sustain the patients condition during the transport till the patient reaches to the emergency and casualty sections of the hospital. The basic ambulatory EHR system is shown as in Figure 7.1.

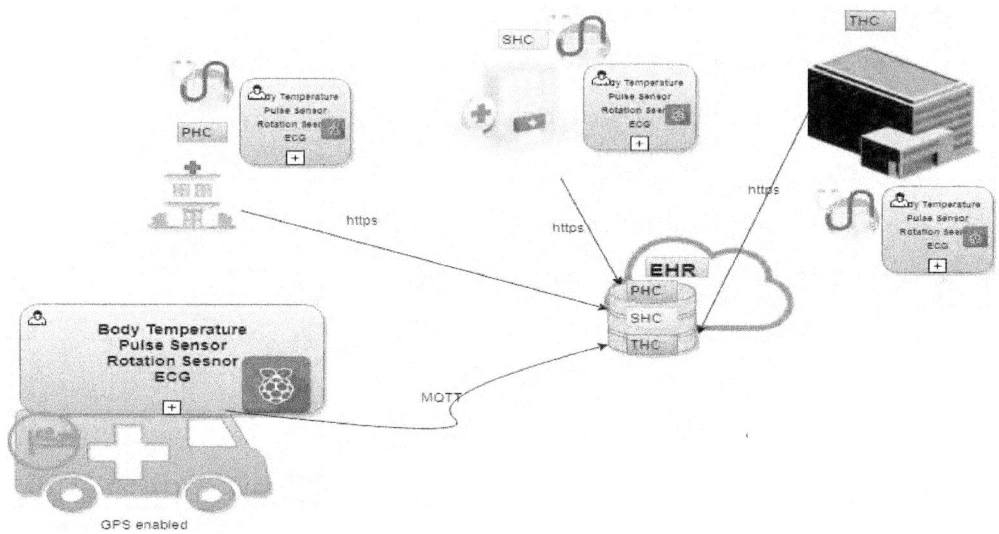

Figure 7.1: Ambulatory EHR

The features to be added further is to capture patient real time data through the sensors including heartbeat pulse sensor, temperature, digital thermometer, accelerometer, and gyroscope sensor attached to the body during the patient shift to the hospital. The sensor data fed into the global cloud database service, using web application healthcare professionals from different levels can continuously monitor the health condition of the patient in real time.

www.ingramcontent.com/pod-product-compliance
Lightning Source LLC
LaVergne TN
LVHW082348200926
848481LV00012B/3500
9798232472382